Anatomy and Exposures of Spinal Nerves

Amgad S. Hanna

Anatomy and Exposures of Spinal Nerves

Illustrations by Mark Ehlers, BS

Amgad S. Hanna
Department of Neurosurgery
University of Wisconsin
Madison, WI
USA

ISBN 978-3-319-14519-8   ISBN 978-3-319-14520-4 (eBook)
DOI 10.1007/978-3-319-14520-4

Library of Congress Control Number: 2015936792

Springer Cham Heidelberg New York Dordrecht London
© Springer International Publishing Switzerland 2015
This work is subject to copyright. All rights are reserved by the Publisher, whether the whole or part of the material is concerned, specifically the rights of translation, reprinting, reuse of illustrations, recitation, broadcasting, reproduction on microfilms or in any other physical way, and transmission or information storage and retrieval, electronic adaptation, computer software, or by similar or dissimilar methodology now known or hereafter developed.
The use of general descriptive names, registered names, trademarks, service marks, etc. in this publication does not imply, even in the absence of a specific statement, that such names are exempt from the relevant protective laws and regulations and therefore free for general use.
The publisher, the authors and the editors are safe to assume that the advice and information in this book are believed to be true and accurate at the date of publication. Neither the publisher nor the authors or the editors give a warranty, express or implied, with respect to the material contained herein or for any errors or omissions that may have been made.

Printed on acid-free paper

Springer International Publishing AG Switzerland is part of Springer Science+Business Media (www.springer.com)

This book is dedicated to my wife Linda for all her support throughout my career
To my parents
To my daughters Barbara, Krista, and Cielle
To my teachers

Foreword

Most physicians' understanding of the anatomy of the spinal or peripheral nerves is limited throughout the fields of medicine, in spite of that system's exquisite importance to normal function. Often this is the result of a lack of experience with the pathology involved or impractical training methods. The functionality of these nerves for qualities of occupation, ambulation, and eye-hand coordination, to name but a few, deserve better. Traditional teaching techniques lack the organization that would enhance understanding and familiarity.

In this volume, Hanna and Ehlers provide a straightforward approach to the surgical anatomy of the spinal nerves by emphasizing the steps of surgery and the actual structures that are seen. Their approach is concise and practical. It emphasizes a surgeon's view of the anatomy in a step-by-step pattern that facilitates understanding of the surrounding musculature, vasculature, and their location and importance. By working through each of the major surgical nerves they cover most clinical scenarios. This makes the book invaluable for students attempting to understand the practical implications of this system or surgeons efficiently studying their work and building on this platform to improve surgical techniques. The practicality of this book is obvious in that it serves the surgeons in an anatomically based fashion through a practical rather than cadaveric approach. It is a welcome addition to literature, and, in a concise fashion, thoroughly covers the topic. The authors are to be congratulated for this contribution.

Robert J. Dempsey, MD, FACS
Department of Neurological Surgery, University of Wisconsin, USA

Peripheral nerve anatomy is often a stumbling block for neurosurgeons preparing for board examinations or contemplating unusual cases. Outside of academic centers, limited volume and sub-specialization in other fields limits the exposure of neurosurgeons to these types of cases. Traditional resources for trainees and those wishing to "brush up" on peripheral nerve cases consist of anatomic textbooks, which largely rely on traditional anatomic exposures to illustrate the structures of interest. These "anatomical" exposures are not always directly relevant to surgical exposures, which in general are less extensile in order to minimize morbidity.

Hanna and Ehlers have provided a concise and useful text for the study of practical peripheral nerve anatomy. This work differs from traditional texts in that the approach is surgically based. The chapters are geared towards answering two or three questions per topic: What anatomy do you see? When do you see it? How do you get there? The text is limited to stripped-down bullet points, which make it an efficient way to review for board examinations or bone up on a case the night before. Ample illustrations and supplemental video aid the learner by providing the same information in multiple formats. This work will be a useful addition to the library of neurosurgeons who perform peripheral nerve cases on occasion and for trainees who are preparing for upcoming cases or for board examination.

Daniel K. Resnick, MD MS
Department of Neurological Surgery, University of Wisconsin, USA

Preface

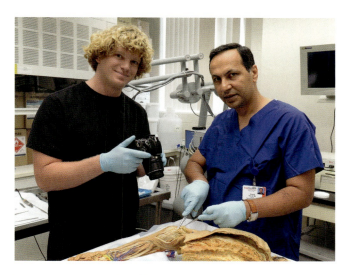

Mark Ehlers (*left, black scrubs*) and Amgad S. Hanna (*right, blue scrubs*) during one of the dissection sessions. This project took about 2 years. Amgad S. Hanna did all the dissections and authored the book; Mark Ehlers did all the video recordings and editing, as well as some photo shoots and editing

This book was mainly inspired by deficiencies in most neurosurgery training programs in the field of peripheral nerves. I prefer the term "spinal nerves" as they contrast to cranial nerves. There is a huge gap between the knowledge requirements and what we actually teach. Simplicity was key in writing this book. I intended to avoid the sophistication of textbooks, and burdening you with lots of references. Each chapter is 1–2 pages of text and plenty of figures. There are video recordings for each described approach. Most of the videos are less than 5 min long. The goal is to be able to review each topic in less than 10 min. Whether you are a resident studying for written boards, a junior faculty studying for oral boards, or getting ready for a case that you haven't seen for years, this book should provide you with easy information, and ample illustrations, concise enough to match your busy schedule. The axial cuts and their MRI correlates are unique and will be extremely helpful not only to the spinal nerve surgeon but also to radiologists. I hope you will enjoy going through the material of this book.

Amgad S. Hanna, MD

Acknowledgments

The author would like to thank:
- Mark Ehlers, BS for all his work in video recordings, editing, as well as photo shoots
- Linda Hanna for reviewing the text for the language content
- R. Shane Tubbs, PhD, PA-C and Edward Bersu for reviewing the scientific content
- Dr Kenneth Lee, MD, musculoskeletal radiologist, Chad Krueger and Caleb Swenson, information processing consultants
- Jeffrey Root, medical media production services
- Robert Schlotthauer and Chad Neuman, body donation program

Contents

Part I Introduction

1 Anatomization: The History of a Uniquely Human Art 3

Part II Upper Body Spinal Nerves

2 Brachial Plexus, Supraclavicular Exposure 11

3 Brachial Plexus, Infraclavicular Exposure 15

4 Brachial Plexus, Combined Supra- and Infraclavicular
   Exposures (Pan-Plexus Exposure) 19

5 Brachial Plexus, Posterior Exposure 21

6 Axillary Nerve 23

7 Radial Nerve 25

8 Musculocutaneous Nerve 29

9 Median Nerve 31

10 Ulnar Nerve 35

11 Suprascapular Nerve 41

12 Spinal Accessory Nerve 45

13 Long Thoracic and Thoracodorsal Nerves 49

14 Intercostal Nerves 51

Part III Lower Body Spinal Nerves

15 Lumbosacral Plexus 55

16 Femoral Nerve 59

17 Lateral Femoral Cutaneous Nerve 63

18 Sciatic Nerve 67

19 Tibial Nerve 71

20 Common Peroneal Nerve (aka Common Fibular Nerve) 75

| 21 | Obturator Nerve | 81 |
| 22 | Ilioinguinal Nerve | 85 |

Part IV Technical Notes

23	Trauma	89
24	Neuroma	95
25	Nerve Sheath Tumors	99
26	Intraneural Ganglion Cysts	103
27	Peripheral Nerve Stimulation	105
28	Dorsal Root Entry Zone (DREZ) Lesion	109

Part V Axial Cuts

| 29 | Upper Extremity Axial Cuts | 115 |
| 30 | Lower Extremity Axial Cuts | 121 |

Index 127

Part I
Introduction

Anatomization: The History of a Uniquely Human Art

The opportunity to "anatomize" the human body. The introduction to a modern gross anatomy class inevitably reminds those newly initiated dissectors of a certain *gravitas* that comes with this opportunity. To introduce this atlas, created with such respect for its object, appreciation for the generosity of those who proffered their body after death for the betterment of our science is almost certainly implicit.

However, there exists a trail through human history, which when followed from its origin in prescientific anatomic exploration through its modern course reminds me of an appreciation that I owe to the anatomists as much as to the anatomized. What started as a search for divinity progressed to a discovery of humanity. What started with *careful observations* by dissectors with *prepared minds* was passed forward as truth. The accumulated truths of human experience define our collective history. We continue this work of discovery, of teaching, and of healing.

While creation stories staunchly carry the Neolithic human's anatomic understanding to us nearly unblemished, as in the story of Adam's rib, the careful observations of many knowledgeable anatomists over centuries have been manipulated, distorted, and overlooked by history [1]. Out of respect, I offer a brief account of what we know about the people who helped humanity see what is on the other side of the skin.

Almost a millennium before the birth of Christ, the Hellenistic Greeks flow into the Aegean Peninsula as Minoan culture ebbs. Collecting and synthesizing the anatomic knowledge of the Mediterranean region, the Greek travelers and traders bring Egyptian and Mesopotamian anatomic experience home with them. While anatomic knowledge manifests in these other Mediterranean cultures by way of spiritual or mystical embarkations, the Greeks' *Canon of Proportion* elegantly incorporates anatomic truths with the meticulous artistic tradition of their Minoan predecessors.

Greece is the patria of our modern anatomic state. The crude descriptions of the wounds of war brought to us through the Greek poet historians, most notably Homer in his *Iliad*, quickly gain sophistication as sculptors demonstrate anatomic understanding at a level that overwhelms language.

While art, philosophy, and medicine flourish as Greece flourishes, a stronghold of knowledge develops across the Mediterranean. Alexandria emerges as a city empowered not by armies or industry but by knowledge. The Alexandrian Library is adorned with the greatest texts of the known world. The halls of the library accommodate the first known scientific dissectors of the human body, Herophilus and Erasistratus, in the fifth century BC [2].

The contribution of Herophilus, the Father of Anatomy, to neuroanatomy is preserved in the names of the *torcular* and the *calamus scriptorius*, both attributed to Herophilus [3]. Erasistratus, the Father of Physiology, described the passage of *animal spirit* from the cerebral ventricles through the intangible lumina of the nerves, engorging and distending muscles to cause contraction. I am unaware of any eponym commemorating Erasistratus.

Though Fathers of Anatomy and Physiology, respectively, the works of these two men are cloaked in mystery and rumor. Their work is known to us second-hand, primarily through the writings of Galen of Pergamon, as the primary texts are burned in two great fires at the Alexandrian Library.

Galen comments on the texts of Herophilus and Erasistratus critically, reworking the understanding of anatomy and physiology with the knowledge he gathers from his own human dissections. Galen described the vertebral column including the most "sacred" bone *sacrum*, described seven cranial nerves, and even described the differences in spinal cord injury based on the injury level with good accuracy [2]. The great cerebral vein is attributed to a worthy Greek.

The contribution of Galen to medicine, anatomy, and philosophy should not be understated. However, he does seem particularly prominent in the history of medicine. His writing is unusually well preserved. His work appealed broadly to pagan and Christian alike; Stoic philosophy infused his writing, his dissection, and his understanding of the world he explored so carefully [3]. His popularity was also aided by his status as the physician to Marcus Aurelius, Commodus, and the gladiators.

Despite his prominence in history, not to mention in Rome itself, Galen does not inspire any notable disciples. The status of physicians in Rome is that of *respected outsider*. The stench of mystery and rumor that surrounds the dissectors in Alexandria follows the early physicians. Stories of death by "anatomization," vivisection, and the like make the field of medicine, rather ironically, the work of those undeserving of Roman citizenry [1, 2].

As the Dark Ages begin, the Roman Empire is crumbling, the Alexandrian library has suffered two great fires, and the mass of accumulated knowledge of the pagan world is consumed or forgotten. The mob of fanatic Christians awaiting the day of reckoning, the day of salvation, and the day of fire and brimstone see plague and pestilence as a sign of apocalyptic imminence. Every day is a reminder that the flesh is of no importance compared to the soul. Anatomic inquiry is of no use and falls by the wayside.

In the ninth century, during the heart of darkness, the seat of human intellect moves East. Arabic translations and commentaries of the great Galenic and Hippocratic works arm the Persian physicians with the tools to advance diagnosis and treatment to a degree that cannot be understated.

Though the neoclassicism of the Renaissance scrubs much of the Arabic dialect from our modern anatomic and medical nomenclature, the names of Rhazes, Hali Abbas, and the great Avicenna are as critical to the story of medicine as Galen, Hippocrates, and Vesalius. However, the care of the body after death in Islamic religion precludes dissection, inhibiting real progress in anatomic discovery during this period. The contribution of Persian physicians is really that of preservation when looking purely at the field of anatomy independent of medicine.

Back in the West, a zeal for scholasticism particularly in the monasteries of Northern Italy inspires the collection and translation of the Greco-Arabic and original Greek texts. This establishes a foundation upon which the first universities are built. At Salerno and Bologna, universities are no more than direct secular extensions of monastic scholasticism.

It is at the University of Bologna that our story accelerates. At the beginning of the fourteenth century, there emerge three students of the university that possess real experience in human dissection, knowledge that does not come from the preserved texts of Greeks or Persians.

The common thread is a mysterious professor, *Thaddeus of Florence*. The work of Thaddeus is largely unknown, but unpublished documents regarding this man are rumored to exist in the Vatican Library [2]. Among his students, *Bartolomeo da Varignana* performs the first documented postmortem examination (which he conducts for forensic purposes at the order of the "jurists" of Bologna's prominent law school), *Henri de Mondeville* brings academic human dissection to the great medical schools that develop in France at Paris and Montpellier, and *Mondino de' Luzzi* pens *Anathomia*, a dissection manual so revolutionary that history labels him as the "Great Restorer of Anatomy" [3].

While there is little more than recapitulation of the Bolognese experience of Mondino and the others for the remainder of the fourteenth century, dissection gains traction in medical school curricula and gains acceptance in popular culture. Even the omnipresent Catholic Church accepts this new mode in the exploration of humanity, owing in large part to Pope Sixtus IV, an alumnus from the University of Bologna. Pope Alexander V, in fact, is examined postmortem prior to his burial at the Church of San Francesco in Bologna, of all places.

Now, the artistic, scientific, and philosophical tides of the time are soon to swell into a great wave of human achievement, the Renaissance.

Here we must acknowledge a great anachronism: Leonardo da Vinci. Although da Vinci's legacy is that of an intellectual superhero, his works do not contribute to the linear advancement of human understanding. When his partner, anatomy professor Marcantonio della Torre, dies in the midst of their creation of an anatomic text, his 700+ anatomic illustrations are shelved. The sketches, though brilliant, are unpublished until 1784.

We return, then, to the work of mere mortals. In the early Renaissance, anatomic knowledge is disseminated in printed versions of Mondino's *Anathomia*, along with illustrated commentaries on Mondino. Anatomy gains popularity in both scientific and lay communities. Anatomic illustrations take on a certain showmanship that suggests a dual purpose: education and entertainment [3].

With the texts of Galen, Mondino, and Mondino's commentators now printed, distributed, and accessible, we have the dénouement of anatomy's modernization. A man not yet 30 years of age named Andreas Vesalius publishes *De humani corporis fabrica*, translated *On the Fabric of the Human Body* (Fig. 1.1). This text has the greatest influence on the course of anatomic science. The *Fabrica*, as it is known, and its sequel *Epitome* are masterpieces that showcase the knowledge, the skill, and the passion of Vesalius [1–3].

It was the accuracy, artistry, and draftsmanship of *Fabrica* that captured the attention and imagination of academics, royalty, and the common person alike (Fig 1.2). The great anatomic dissection and illustration of Vesalius not only conveyed the facts hidden under the skin but also evoked the response that all great art evokes [4–6]. A dramatized and artistic presentation of anatomy persisted despite evolution in the processes of dissection, illustration, and publication. A plate from the seventeenth-century *Anatomia humani corporis* by the Dutch anatomist Govard

Fig. 1.1 Frontispiece from Vesalius' masterpiece *On the Fabric of the Human Body*, 1543, known fondly as *Fabrica*. Vesalius is the young, bearded man boldly instructing the clamoring crowd, not from the professor's chair as was tradition but from the work itself, the careful anatomization of the human body [5] (Courtesy of Ebling Library, Rare Books and Special Collections, University of Wisconsin-Madison)

Bidloo (Fig 1.3) demonstrates the artistic and dramatic presentation of anatomy, with a prosected arm dangling from the dissection table, muscles pinned back as the figure was sketched [4, 7].

Like da Vinci, like Galen, and like Herophilus, Vesalius was not divine. These men might not have thought that what they were doing would change the world like it has. The greatness of their work can be explained without mysticism, divine intervention, or even luck. It was by the power of *careful observation* and a *prepared mind* that these anatomists changed the collective understanding of the human body.

We are learning about the nervous system now like centuries past have learned about the rest of the body. Remember the work of the functional anatomists of the nineteenth century: Charcot, Bell, and Magendie. Remember the work of the surgical anatomists of the twentieth century: Cushing, Penfield, and Rhoton. How will the twenty-first century anat-

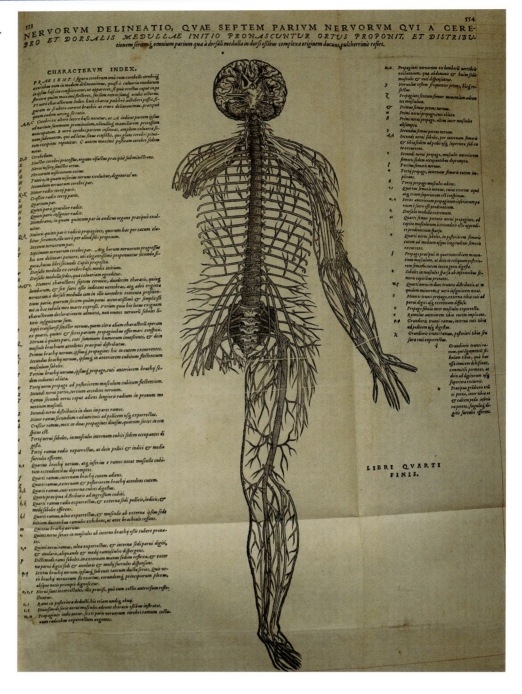

Fig. 1.2 In this foldout figure from the 1555 edition of *Fabrica*, Vesalius delineates the nervous system impressively. Although dissections and sketches had to be carried out hastily in the absence of tissue fixation techniques, the accuracy of Vesalius's work set him apart from his contemporaries. The brachial plexus and lumbosacral plexus, despite some errors, are fairly represented [6] (Courtesy of Ebling Library, Rare Books and Special Collections, University of Wisconsin-Madison)

omist be remembered? Our work is influenced by our heritage, our environment, our way of thinking, our educational background, our rivals, and our collaborators.

By looking at the accumulated achievements of the past, perhaps it is easy to say that everything that can be discovered has already been discovered. But I argue that this well will not run dry. The accumulated knowledge of humanity is moving from the monotonous shelves of the world's libraries to the phone in my back pocket. The time for discovery is now.

With careful observation, this atlas has been compiled. It has been created so that you might draw forth from the well of knowledge merely one Pail among all that you will ever learn. When you imbibe, the knowledge will manifest uniquely. No student will have the same preface, and no teacher will have the same conclusion.

With this collection of anatomic truths drawn forth from under the skin of the laudable dissected, prepare your mind *to discover*, *to teach*, and *to heal* as no one has done before you.

Fig. 1.3 The musculocutaneous nerve piercing the coracobrachialis in Bidloo, 1685. Anatomic dissection and illustration proceeded in the traditions of Vesalius for centuries. Although printing techniques and anatomic knowledge pressed forward, the stylized and dramatized artistic presentation of anatomic dissection was preserved. Vesalius' work was printed with wooden plates, which were carved by hand as a "negative" image and then inked and pressed. The plate shown here from Bidloo's text was etched in copper plates, allowing the texture and lines of the anatomy to be printed in more fine detail. The craftsmanship required to print these Renaissance texts remains impressive today [4, 7] (Courtesy of Ebling Library, Rare Books and Special Collections, University of Wisconsin-Madison)

Christopher D. Baggott
Madison, WI, USA

References

1. Hurren ET (2014) Personal communication. (CD Baggott, Interviewer)
2. Singer C (1957) A short history of anatomy from the Greeks to Harvey. Dover Publications, New York
3. Singer C (1982) Evolution of anatomy : a short history of anatomical and physiological discovery to Harvey : being the substance of the Fitzpatrick lectures delivered at the Royal College of Physicians of London in the years 1923 and 1924. Second Ed. Alfred A. Knopf, New York (Hertford: Stephen Austin & Sons)
4. Fowler-Sullivan M (2014) Personal communication. (CD Baggott, Interviewer)
5. Vesalius A, van Calcar JS (1543) De humani corporis fabrica libri septem. ex officina Joannis Oporini, Basel
6. Vesalius A, Joannes O (1555) De humani corporis fabrica libri septem. Per Ioannem Oporinum, Basel
7. Bidloo G, de Lairesse G (1685) Anatomia Hvmani Corporis, Centum & Quinque Tabvlis, Per Artificiosiss. sumptibus viduae Joannis à Someren [etc.], Amsteledami

Part II
Upper Body Spinal Nerves

Brachial Plexus, Supraclavicular Exposure

Surgical Anatomy

The supraclavicular brachial plexus includes the roots and the trunks. The brachial plexus is usually formed by the ventral rami of the spinal nerves from C5 to T1. Contribution is very common from C4 (prefixed) and uncommon from T2 (postfixed) [1]. It forms between the scalenus anterior and scalenus medius muscles and lies in the posterior triangle of the neck (bound by the sternocleidomastoid [SCM] anteriorly, the trapezius posteriorly, and the clavicle caudally). C5 and C6 join to make the upper trunk, C7 continues as the middle trunk, and C8 and T1 join to make the lower trunk. The suprascapular nerve (C5, C6) is the largest branch at the root/trunk level. It arises from the cranial aspect of the upper trunk, close to the divisions. The dorsal scapular nerve (nerve to rhomboids and levator scapulae) comes off the C5 ventral ramus; the long thoracic nerve (nerve to serratus anterior) comes off the C5, C6, and occasionally C7 ventral rami and usually runs through the scalenus medius muscle. The nerve to subclavius (C5, C6) comes off the upper trunk.

Exposure

1. Position: supine, with the back of the table up and head turned contralaterally, with or without a shoulder bump (Fig. 2.1a).
2. Incision: oblique along the posterior border of the SCM muscle or preferably horizontal curvilinear along or parallel to a skin crease (Fig. 2.1b).
3. The skin flaps are elevated between the subcutaneous fat and platysma muscle using DeBakey forceps and Metzenbaum scissors, while the assistant is retracting the skin using two Senn retractors.
4. The platysma is opened along its fibers and undermined (Fig. 2.2).
5. The posterior border of the SCM is identified. Care should be taken not to injure the great auricular nerve (which winds around the posterior border of the SCM) or the spinal accessory nerve cranial to it.
6. The fat pad is then dissected free from the posterior border of the SCM. The omohyoid muscle is found deep to the SCM and lies horizontally (Fig. 2.3a); it is tagged with 2-0 Ethibond sutures, cut and then retracted (Fig. 2.3b). Cutaneous branches of the cervical plexus are usually encountered and may have to be sacrificed.
7. The transverse cervical and suprascapular vessels are also encountered and need to be ligated with 2-0 silk ties and then cut. They run horizontally in the caudal part of the exposure, sometimes behind the clavicle (Fig. 2.3c).
8. The scalenus anterior (anticus) muscle is then identified with the phrenic nerve anterior to it from cranio-lateral to caudo-medial (Fig. 2.4a). The nerve can also be tested by electrical stimulation (Fig. 2.4b).
9. The phrenic nerve followed cranially will lead to the C5 ventral ramus; the latter followed laterally will join C6 to form the upper trunk.
10. The middle trunk formed by C7 is found caudal to the upper trunk and in a deeper plane.
11. Caudal to the middle trunk is the subclavian artery and posterior to it is the lower trunk (Fig. 2.5a). Following the lower trunk medially leads to the C8 nerve root. The latter is usually in line with the lower trunk. T1 ventral ramus ascends to meet C8 at an angle to form the lower trunk (Fig. 2.5b). T1 is sometimes easier to find by retracting the lower trunk anteriorly.
12. The trunks usually divide behind the clavicle. The suprascapular nerve, branching off the upper trunk, is usually encountered at the caudal part of the supraclavicular exposure.
13. The spinal accessory nerve, usually intact in brachial plexus injuries and used as a donor for nerve transfers, is

Electronic supplementary material The online version of this chapter 10.1007/978-3-319-14520-4_2 contains supplementary material, which is available to authorized users.

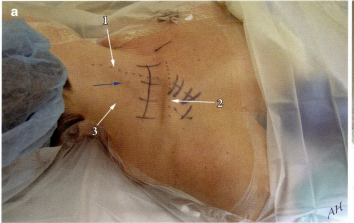
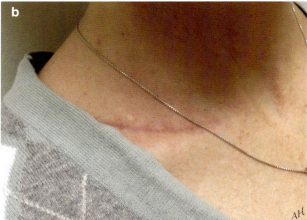

Fig. 2.1 Right supraclavicular brachial plexus exposure. (**a**) The patient is positioned supine with the head turned contralaterally; the back of the bed is elevated, with or without a shoulder bump. *White arrows* show the boundaries of the posterior triangle of the neck: *1* posterior border of sternocleidomastoid (SCM), *2* clavicle, *3* anterior border of the trapezius. *Blue arrow*: external jugular vein. Incision can be made along the posterior border of the SCM or, as depicted here, horizontal along or parallel to a skin crease. The latter is usually more aesthetic (**b**)

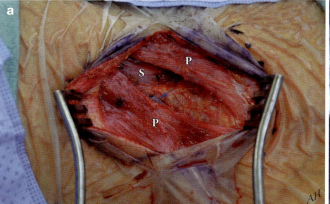
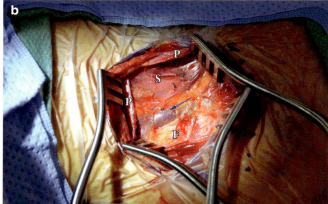

Fig. 2.2 The platysma (*P*) is divided along its fibers (**a**) and retracted (**b**). This reveals the posterior border of the SCM (*S*), external jugular vein (*blue arrow*), and fat pad (*F*)

found in the fat pad under the anterior border of the trapezius muscle (see also Chap. 12).

Complications

1. Diaphragmatic paralysis from phrenic nerve injury or use as a donor for nerve transfers. Usually well tolerated but can be complicated by reduced pulmonary vital capacity. However, this is rarely symptomatic.
2. Vascular injuries to the subclavian artery, internal jugular vein, transverse cervical vessels, or suprascapular vessels. Small vessels can be ligated or clipped. Large vessels require repair sometimes by a vascular surgeon. Vertebral artery injuries could be complicated by bleeding or stroke. Care should be taken if dissection is to be carried too proximally towards the foramina.
3. Thoracic duct injury, on the left side. Direct repair with nonabsorbable suture or patching with fat pad and supplementing with fibrin glue.
4. Pneumothorax, from injury to the pleura and possibly the lung apex. The pleura can be directly repaired. Chest tube may or may not be needed. Postoperative chest X-ray should be obtained.
5. Iatrogenic nerve injuries.
6. Wound infection.

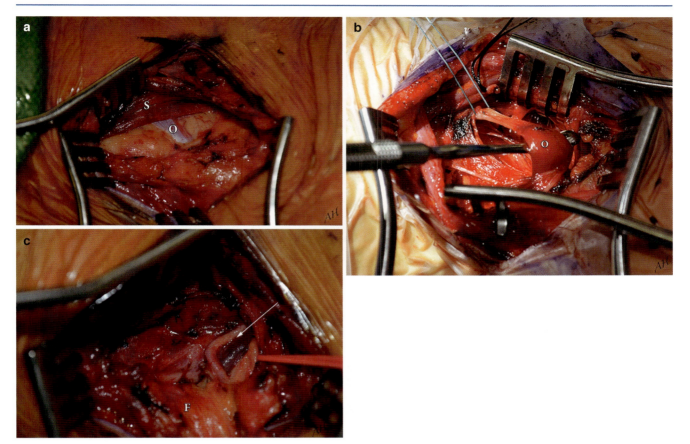

Fig. 2.3 (**a**) The omohyoid muscle (*O*) is identified deep to the SCM (*S*), tagged, and cut (**b**). (**c**) The supraclavicular pad of fat (*F*) is dissected and retracted laterally. The suprascapular and transverse cervical vessels (*arrow*) are ligated and cut

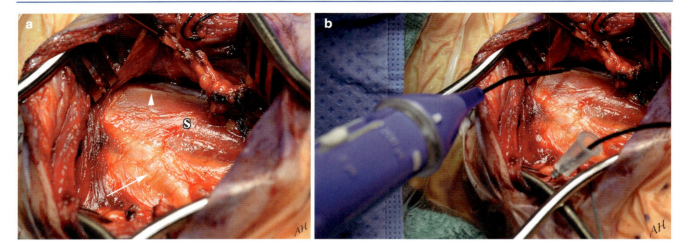

Fig. 2.4 (**a**) The phrenic nerve (*arrowhead*) is observed on the surface of the scalenus anterior muscle (*s*). It can be identified by electrical stimulation (**b**). Note the upper trunk (*arrow*) seen through the fat

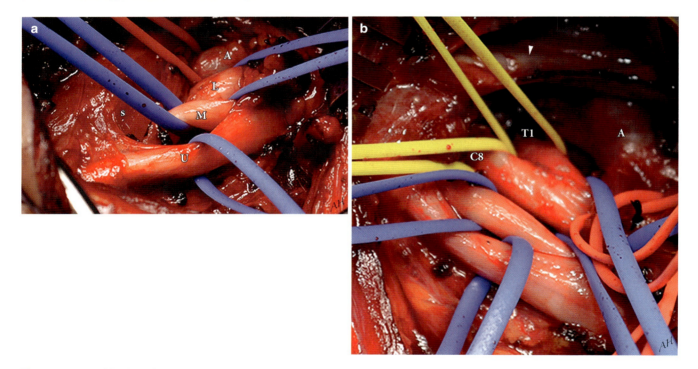

Fig. 2.5 (**a**) Identification of the trunks of the brachial plexus. *U* upper trunk (C5, 6), *M* middle trunk (C7), *L* lower trunk (C8, T1), *A* subclavian artery, *s* scalenus anterior. (**b**) More medial exposure (thoracic outlet decompression) reveals the C8 and T1 ventrally. *Arrowhead*: phrenic nerve

Reference

1. Pellerin M, Kimball Z, Tubbs RS, Nguyen S, Matusz P, Cohen-Gadol AA, Loukas M (2010) The prefixed and postfixed brachial plexus: a review with surgical implications. Surg Radiol Anat 32(3):251–260

Brachial Plexus, Infraclavicular Exposure

Surgical Anatomy

The infraclavicular brachial plexus includes the cords and the terminal branches. They correlate closely with the axillary artery. The posterior cord is formed by the junction of the posterior divisions of the three trunks; it first runs lateral and then posterior to the axillary artery. It gives off the thoracodorsal nerve (to latissimus dorsi muscle) and the upper (to subscapularis) and lower subscapular nerves (to subscapularis and teres major muscles). Terminal branches include the axillary and radial nerves. The lateral cord forms from the junction of the anterior divisions of the upper and middle trunks; it first runs anterior and then lateral to the axillary artery. It gives off the lateral pectoral nerves. Terminal branches include the musculocutaneous nerve and the lateral contribution to the median nerve. The medial cord is the continuation of the anterior division of the lower trunk; it first runs posterior and then medial to the axillary artery. It gives off the medial pectoral nerve, medial brachial cutaneous, and medial antebrachial cutaneous nerves. Terminal branches include the ulnar nerve and the medial contribution to the median nerve.

Anatomical Variations

Communications between lateral and medial cords may occur [1]. More than one contribution to the median nerve can be observed especially from the lateral cord. The lateral pectoral nerve may arise more proximally from the anterior division of the upper trunk or much less frequently from the anterior division of the middle trunk [2].

Electronic supplementary material The online version of this chapter 10.1007/978-3-319-14520-4_3 contains supplementary material, which is available to authorized users.

Exposure

1. Position: supine, with the back of the table up and the patient's head turned contralaterally.
2. Incision is made along the deltopectoral groove (Fig. 3.1).
3. The deltoid is separated from the pectoralis major muscle. In this interval, the cephalic vein is usually encountered (Fig. 3.2). It is important to preserve the vessels in this area for a potential free muscle flap.
4. The coracoid process is then palpated. Pectoralis minor inserts into the medial part of the coracoid process, while the short head of the biceps brachii and coracobrachialis originate from its lateral part (Fig. 3.3). The pectoralis minor is tagged with 2-0 Ethibond sutures, transected, and then retracted. Underneath the pectoralis minor is a fatty layer that contains the brachial plexus.
5. Dissection through this fatty layer uncovers the lateral cord as it gives off the musculocutaneous nerve (Fig. 3.4) and the lateral contribution to the median nerve.
6. Posterolateral to the lateral cord, the posterior cord can be found as it gives off the axillary and radial nerves.
7. The axillary artery is then found medial and posterior to the lateral cord and partially covers the medial cord.
8. The medial cord is located medial and posterior to the axillary artery. It gives off the ulnar nerve and medial contribution to the median nerve. Medial to the ulnar nerve are the medial antebrachial cutaneous and medial brachial cutaneous nerves (Figs. 3.5, 3.6, and 3.7).
9. The terminal branches of the medial and lateral cords form an "M" and "hug" the axillary artery (Fig. 3.8).

Complications

1. Vascular injuries
2. Pneumothorax
3. Iatrogenic nerve injuries
4. Wound infection

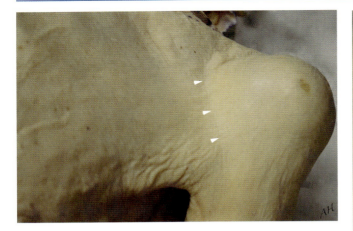

Fig. 3.1 Left infraclavicular brachial plexus exposure. The incision is made along the deltopectoral groove (*arrowheads*)

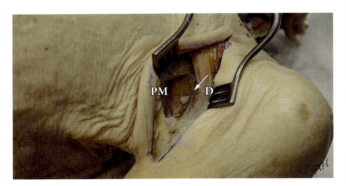

Fig. 3.2 The deltoid (*D*) and pectoralis major (*PM*) are retracted. The cephalic vein (*arrow*) is observed in between

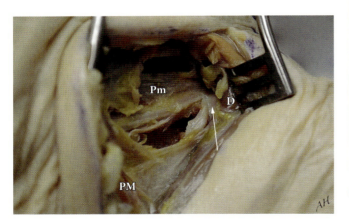

Fig. 3.3 Pectoralis minor (*Pm*) is found, tagged, cut, and retracted. Note the common tendon of the short head of the biceps brachii and coracobrachialis (*arrow*) laterally. *PM* pectoralis major, *D* deltoid

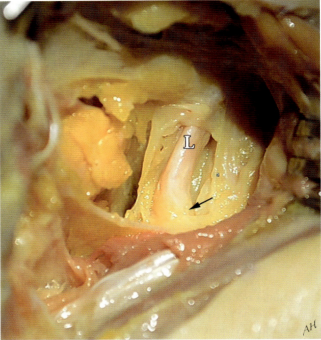

Fig. 3.4 The pad of fat deep to the pectoralis minor contains the brachial plexus. The lateral cord (*L*) is first encountered as it gives off the musculocutaneous nerve (*arrow*)

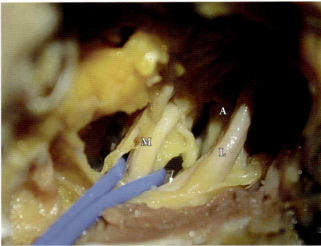

Fig. 3.5 Further dissection reveals the medial cord (*M*). The lateral cord (L) gives off the lateral contribution to the median nerve (*arrow*). *A* axillary artery

Complications

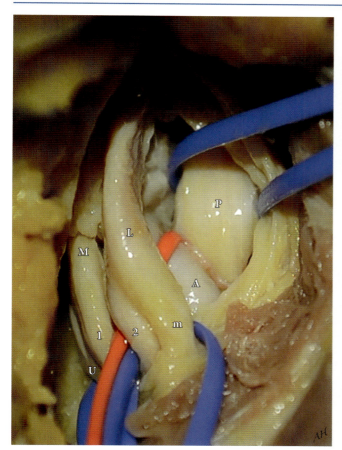

Fig. 3.6 The three cords are exposed. *M* medial cord, *L* lateral cord, *P* posterior cord, *A* axillary artery, *U* ulnar nerve, *m* musculocutaneous nerve, *1* medial contribution to median nerve, *2* lateral contribution to median nerve

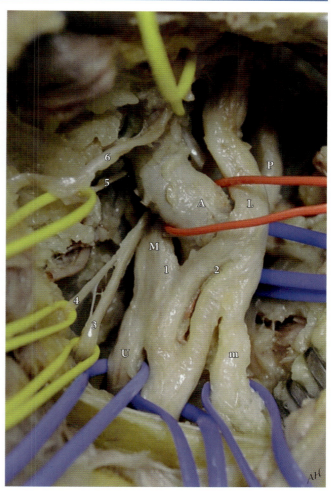

Fig. 3.7 The main branches of the medial (*M*) and lateral (*L*) cords are exposed. From the medial cord: medial contribution to median nerve (*1*), ulnar nerve (*U*), medial antebrachial cutaneous nerve (*3*), medial brachial cutaneous nerve (*4*), and medial pectoral nerve (*5*). From the lateral cord: lateral contribution to median nerve, here duplicated (*2*), musculocutaneous nerve (*m*), and lateral pectoral nerve (*6*). Note here that the lateral and medial pectoral nerves join. *P* posterior cord, *A* axillary artery

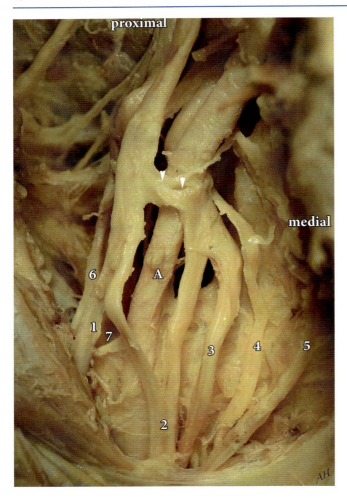

Fig. 3.8 Right infraclavicular exposure showing the components of the "M": *1* musculocutaneous nerve, *2* median nerve, *3* ulnar nerve. *A* Axillary artery, *4* medial antebrachial cutaneous nerve, *5* medial brachial cutaneous nerve, *6* axillary nerve, *7* radial nerve. Note the communication between the medial and lateral cords (*arrowheads*)

References

1. Goel S, Rustagi SM, Kumar A, Mehta V, Suri RK (2014) Multiple unilateral variations in medial and lateral cords of brachial plexus and their branches. Anat Cell Biol 47:77–80
2. Arad E, Li Z, Sitzman T, Agur A, Clarke H (2014) Anatomic sites of origin of the suprascapular and lateral pectoral nerves within the brachial plexus. Plastic and reconstructive surgery 133(1): 20e–27e

Brachial Plexus, Combined Supra- and Infraclavicular Exposures (Pan-Plexus Exposure)

Surgical Anatomy

The divisions of the three trunks are typically located behind the clavicle.

Exposure [1]

1. Supraclavicular exposure as in Chap. 2.
2. Infraclavicular exposure as in Chap. 3.
3. Combined exposure: the lateral end of the supraclavicular incision is connected to the cranial end of the infraclavicular incision.
4. The clavicle is identified and surrounded by a lap sponge or a Penrose drain to allow retraction up and down (Fig. 4.1).
5. The posterior divisions of the three trunks form the posterior cord (Figs. 4.2 and 4.3).
6. The anterior divisions of the upper and middle trunks join to make the lateral cord.
7. The anterior division of the lower trunk continues as the medial cord.
8. The upper trunk terminal branches are arranged from craniodorsal to caudoventral as follows: suprascapular nerve, posterior division, and then anterior division (Fig. 4.4).

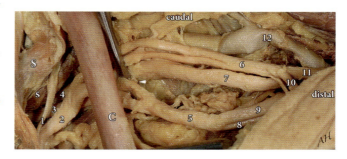

Fig. 4.1 For combined right supra- and infraclavicular exposures, the clavicle (*C*) is exposed and retracted cranially or caudally as needed. *S* sternocleidomastoid, *s* scalenus anterior, *1* phrenic nerve, *2* upper trunk, *3* middle trunk, *4* lower trunk, *5* posterior cord made by the junction of the posterior divisions of the three trunks. A small component (*arrowhead*) from the lower trunk joins the posterior division of the middle trunk, and then they join the posterior division of the upper trunk; *6* lateral cord, made by the junction between the anterior divisions of the upper and middle trunks, *7* medial cord, made by the continuation of the anterior division of the lower trunk, *8* axillary nerve, *9* radial nerve, *10* musculocutaneous nerve, *11* lateral contribution to median nerve, *12* axillary artery

Fig. 4.2 Other branches of the same right brachial plexus are identified. From the upper trunk, *1* suprascapular nerve. From the posterior cord: *2* thoracodorsal nerve, *3* axillary nerve, *4* radial nerve. From the lateral cord: *5* musculocutaneous nerve, *6* lateral contribution to the median nerve, *7* lateral pectoral nerve. From the medial cord: *8* medial pectoral nerve, *9* medial brachial cutaneous nerve, *10* medial antebrachial cutaneous nerve, *11* ulnar nerve, *12* medial contribution to the median nerve. *13* axillary artery, *14* median nerve

Electronic supplementary material The online version of this chapter 10.1007/978-3-319-14520-4_4 contains supplementary material, which is available to authorized users.

Reference

1. Tse R, Pondaag W, Malessy M (2014) Exposure of the retroclavicular brachial plexus by clavicle suspension for birth brachial plexus palsy. Tech Hand Up Extrem Surg 18(2):85–88

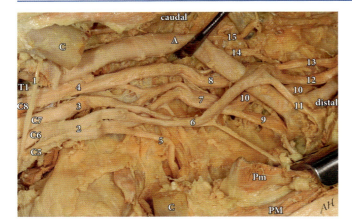

Fig. 4.3 Overview of the right brachial plexus from a different specimen after division of the clavicle (C) for illustration. *PM* pectoralis major, *Pm* pectoralis minor, *A* artery, subclavian → axillary; *1* phrenic nerve, *u*pper trunk, *3* middle trunk, *4* lower trunk, *5* suprascapular nerve, *6* lateral cord, *7* posterior cord, *8* medial cord, *9* axillary nerve, *10* radial nerve, *11* musculocutaneous nerve, *12* median nerve, *13* ulnar nerve, *14* medial antebrachial cutaneous nerve, *15* medial brachial cutaneous nerve

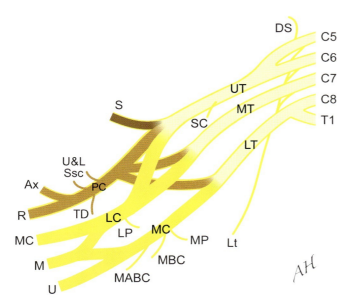

Fig. 4.4 Diagrammatic representation of the right brachial plexus. Note the arrangement of the branches of the upper trunk from cranial to caudal: suprascapular nerve (*S*), posterior division, and then anterior division. *DS* dorsal scapular nerve, *Lt* long thoracic nerve, *UT* upper trunk, *MT* middle trunk, *LT* lower trunk, SC nerve to the subclavius, *PC* posterior cord, *LC* lateral cord, *MC* medial cord, *LP* lateral pectoral nerve, *MP* medial pectoral nerve, *MBC* medial brachial cutaneous nerve, *MABC* medial antebrachial cutaneous nerve, *U & L Ssc* upper and lower subscapular nerves, *TD* thoracodorsal nerve, *Ax* axillary nerve, *R* radial nerve, *MC* musculocutaneous nerve, *M* median nerve, *U* ulnar nerve

Brachial Plexus, Posterior Exposure

Surgical Anatomy

From a posterior approach, the brachial plexus is anterior to the following muscle planes from superficial to deep: trapezius, rhomboids and levator scapulae, serratus posterior superior, scalenus posterior, and scalenus medius. In this exposure, the subclavian artery is anterior to (deep to) the lower trunk.

Exposure [1]

1. Position: prone on chest rolls.
2. Incision: paramedian, between the midline and medial border of the scapula (Fig. 5.1).
3. The trapezius (Fig. 5.2) is incised vertically.
4. The rhomboids and levator scapulae (Fig. 5.3) are incised vertically.
5. This allows retraction of the scapula laterally.
6. The serratus posterior superior (Fig. 5.4) and posterior and middle scalene muscles are partially resected and detached from the first rib. Care should be taken not to mistake the second rib for the first rib. Always palpate above and deep to the most cranial rib to make sure it is indeed the first rib.
7. Using a periosteal elevator (Doyen), the first rib is dissected taking care not to injure the neurovascular bundle caudal to it, the subclavian vessels cranial to it, or the parietal pleura deep to it.
8. The first rib is then resected using a rib cutter or Leksell rongeur.
9. This should expose the trunks.
10. These can be followed proximally to nerve roots, where a bony foraminotomy can be performed using Kerrison rongeurs. Venous bleeding is usually encountered in the foramen and can easily be tamponade with Gelfoam and thrombin.
11. The trunks can be followed distally to divisions (Fig. 5.5).
12. The suprascapular nerve can be identified coming off the upper trunk going towards the suprascapular notch.
13. The long thoracic nerve arises from the posterior aspect of the upper ventral rami (C5–C7), within scalenus medius, and travels towards the serratus anterior outer surface.

Fig. 5.1 Posterior exposure of the right brachial plexus. Position: prone. Incision: paramedian, between the midline and medial border of the scapula

Fig. 5.2 The trapezius (*T*) is exposed and incised

Electronic supplementary material The online version of this chapter 10.1007/978-3-319-14520-4_5 contains supplementary material, which is available to authorized users.

A.S. Hanna, *Anatomy and Exposures of Spinal Nerves*,
DOI 10.1007/978-3-319-14520-4_5, © Springer International Publishing Switzerland 2015

Fig. 5.3 The rhomboids (*R*) and levator scapulae (*L*) are exposed and incised

Fig. 5.4 The serratus posterior superior (*S*) is exposed and incised, thus exposing the ribs (*r*)

14. The spinal accessory nerve can be found on the undersurface of the trapezius muscle towards its anterior border.

Complications

1. Pneumothorax
2. Injury to the neurovascular bundle under the first rib
3. Vertebral artery injury
4. Subclavian artery injury
5. Wound infection
6. Iatrogenic nerve injury

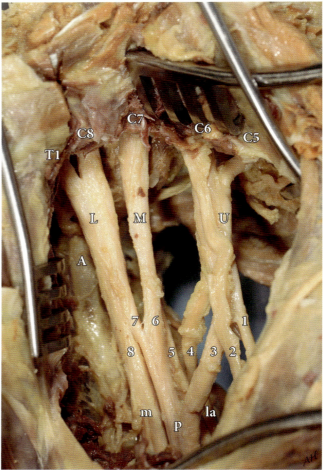

Fig. 5.5 Posterior view of the brachial plexus after resection of the first rib and posterior and middle scalenes. *U* upper trunk, *M* middle trunk, *L* lower trunk, *A* subclavian artery, *la* lateral cord, *p* posterior cord, *m* medial cord, *1* suprascapular nerve, *2* long thoracic nerve, *3* upper trunk, posterior division, *4* upper trunk, anterior division, *5* middle trunk, anterior division, *6* middle trunk, posterior division, *7* lower trunk, posterior division, *8* lower trunk, anterior division

Reference

1. Hager S, Backus TC, Futterman B, Solounias N, Mihlbachler MC (2014) Posterior subscapular dissection: an improved approach to the brachial plexus for human anatomy students. Ann Anat 196(2–3):88–91

Axillary Nerve

Surgical Anatomy

The axillary nerve (C5 and C6) comes off the posterior cord at the level of the coracoid process [1]. Fibers originate mainly from the posterior division of the upper trunk. It lies posterior to the axillary artery and then the musculocutaneous nerve. It leaves the axilla to enter the posterior aspect of the upper arm through the quadrangular space, accompanied by the posterior circumflex humeral vessels. The quadrangular space is bound by the subscapularis and teres minor cranially, teres major caudally, long head of triceps medially, and humerus laterally. Teres major separates the axillary nerve from the radial nerve. The axillary nerve supplies the teres minor and the deltoid muscles. It divides into: an anterior division that is deeper and in closer relationship to the humerus, and a posterior division that gives off the upper lateral cutaneous nerve of the arm.

Clinical Significance

The axillary nerve is frequently injured with shoulder dislocations or humeral neck fractures. This causes inability to abduct the arm.

Exposure

Anterior Approach

1. Infraclavicular exposure as in Chap. 3.
2. The axillary nerve is identified as it comes off the posterior cord at the level of the coracoid process (Fig. 6.1).

Electronic supplementary material The online version of this chapter 10.1007/978-3-319-14520-4_6 contains supplementary material, which is available to authorized users.

Fig. 6.1 Origins of the axillary and radial nerves from the posterior cord of the left brachial plexus. *1* scalenus medius, *2* scalenus anterior, *3* phrenic nerve, *4* upper trunk, *5* nerve to the subclavius, *6* suprascapular nerve, *7* upper trunk, posterior division, *8* upper trunk, anterior division, *9* middle trunk, *10* lower trunk, *11* middle trunk, posterior division, *12* middle trunk, anterior division, *13* lower trunk, posterior division, *14* lower trunk, anterior division → medial cord, *15* lateral cord, *16* posterior cord, *17* upper and lower subscapular nerves, *18* axillary nerve, *19* radial nerve, *20* musculocutaneous nerve, *21* early coracobrachialis branch, *22* median nerve, lateral root, *23* subclavian artery

3. It is followed laterally as it enters the quadrangular space with the posterior circumflex humeral vessels.
4. The axillary nerve then enters the posterior aspect of the arm.

Posterior Approach (Fig. 6.2)

1. The patient can be positioned supine with a bump under the shoulder, lateral, or prone.
2. Incision is made in the posterior aspect of the arm along the posterior border of the deltoid muscle.
3. The incision is deepened through the superficial and deep fasciae.
4. The posterior border of the deltoid muscle is identified with the cutaneous branch winding around from deep to superficial.

5. This branch is important to identify. It is then followed deeply under the deltoid to find its origin from the posterior division of the axillary nerve.
6. The posterior division can be followed proximally to find the main stem of the axillary nerve coming out of the quadrangular space and giving off also the anterior division after a very short course. The anterior division is very deep and in close relationship to the humerus.

Reference

1. Kang MS, Woo JS, Hur MS, Lee KS (2014) Spinal nerve compositions and innervations of the axillary nerve. Muscle Nerve 50(5):856–858

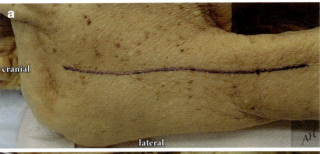
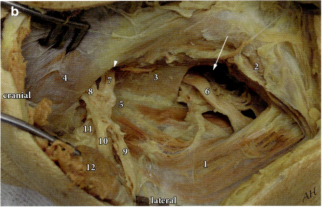

Fig. 6.2 Left axillary and radial nerves exposure in the posterior aspect of the arm. (**a**) Position: prone, lateral, or supine with a shoulder bump. Incision, along the posterior border of the deltoid, extends distally between the long and lateral heads of the triceps. (**b**) The radial nerve (*6*) is exposed between the lateral (*1*) and long (*2*) heads of the triceps and then followed proximally till the lower triangular space (triangular interval) (*arrow*). To expose the axillary nerve, the upper lateral cutaneous nerve of the arm (*9*) is first encountered around the posterior border of the deltoid (*12*); followed proximally, it will lead to the posterior division (*10*) of the axillary nerve; and this will lead to the main trunk of the axillary nerve (*7*) towards the quadrangular space (*arrowhead*). The anterior division (*11*) of the axillary nerve is deep towards the humerus (*5*). The nerve (*8*) to teres minor (*4*) can also be dissected. Teres major (*3*) separates the radial nerve from the axillary nerve

Radial Nerve

Surgical Anatomy

The radial nerve (C5, C6, C7, C8, and T1) arises from the posterior cord at the level of the coracoid process, posterior to the axillary artery and lateral cord. It exits the axilla towards the posterior arm through the lower triangular space (triangular interval) bound by the teres major cranially, the long head of triceps medially, and the humerus laterally. The triceps branches come off the radial nerve high in the axilla. The nerve then courses in the posterior arm, in the spiral groove along with the profunda brachii vessels. It then enters the anterior compartment of the arm between the biceps brachii and triceps and then between the biceps brachii and brachioradialis. It gives off the brachioradialis, extensor carpi radialis longus and brevis (ECRL and ECRB) branches, and the posterior interosseous nerve (PIN) and then continues as the superficial radial nerve (pure cutaneous). The PIN courses through the supinator muscle (arcade of Fröhse) and supplies the remaining muscles of the posterior forearm. The superficial radial nerve emerges from under the brachioradialis tendon and supplies the skin of the dorsum of the hand laterally as well as the dorsum of the lateral three and a half fingers up to the middle phalanx.

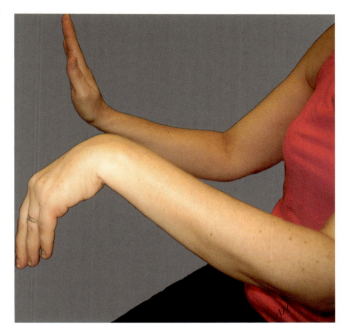

Fig. 7.1 Left wrist drop from radial nerve palsy

Clinical Significance

- Radial nerve injury in the spiral groove spares the triceps branches and results in wrist drop (Fig. 7.1). Sensory loss is limited to the skin overlying the anatomical snuffbox. Injury could occur with humeral shaft fractures or during their surgical plating.
- PIN injury spares ECRL and ECRB branches resulting in wrist extension with radial deviation (Fig. 7.2), finger drop, and no loss of cutaneous sensation. Compression can occur at the arcade of Fröhse [1] or the leash of Henry (vascular) [2].

Exposure

Anterior Approach, Axilla (See Fig. 6.1)

1. Infraclavicular exposure as in Chap. 3.
2. The radial nerve is identified at the level of the coracoid process as it comes off the posterior cord. It is deep to the lateral cord and axillary artery.
3. It is followed distally as it enters the lower triangular space with the profunda brachii vessels.
4. The radial nerve then enters the posterior aspect of the arm.

Electronic supplementary material The online version of this chapter 10.1007/978-3-319-14520-4_7 contains supplementary material, which is available to authorized users.

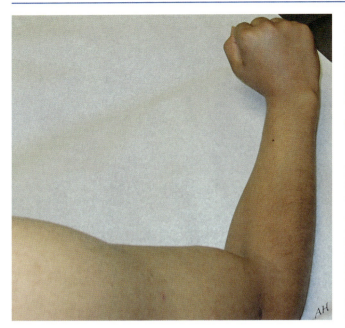

Fig. 7.2 Right posterior interosseous nerve (PIN) palsy. Note the radial deviation of the wrist when the patient attempts to extend it

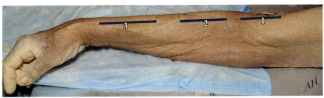

Fig. 7.3 Incisions for radial nerve exposures in the distal arm (*1*), proximal (*2*), and distal forearm (superficial radial) (*3*)

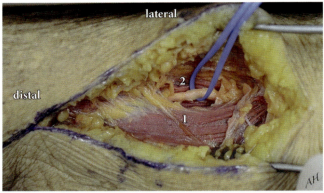

Fig. 7.4 In the distal arm, the radial nerve (*blue loop*) is found between the biceps brachii (*1*) medially and the triceps (*2*) laterally

Posterior Arm (See Fig. 6.2)

1. The patient can be positioned supine with a bump under the shoulder, lateral, or prone.
2. Incision is made in the posterior aspect of the arm along the posterior border of the deltoid and then extended distally in the mid-posterior arm.
3. The incision is deepened through the superficial and deep fasciae.
4. The raphe between the long and lateral heads of the triceps is then identified.
5. The raphe is opened and the two heads of triceps are separated.
6. This reveals the radial nerve and its different triceps branches.
7. It can be followed more distally between the lateral and medial heads of the triceps.

Anterior Elbow (Figs. 7.3 and 7.4)

1. Position: supine, with the arm abducted.
2. Incision is made in the anterior aspect of the lower arm between the biceps brachii medially and the triceps and then brachioradialis laterally.
3. The superficial and deep fasciae are opened.
4. The radial nerve can be found in the interval between the biceps brachii medially and the triceps laterally.
5. Followed distally, the nerve is between the biceps brachii medially and the brachioradialis laterally.

Proximal Forearm (Fig. 7.5)

1. Position: supine, with the arm abducted.
2. Incision is made along the anterior border of the brachioradialis muscle or between brachioradialis and ECRL.
3. The radial nerve can be found between the brachioradialis laterally and the biceps brachii and brachialis medially.
4. Followed distally, we can identify the branches to brachioradialis, ECRL, and ECRB.
5. The PIN comes off the radial nerve and enters the posterior forearm under the aponeurotic edge of the supinator (arcade of Fröhse). Vessels could come in close relationship with the PIN in this location (leash of Henry).

Distal Forearm, Superficial Radial Nerve
(Fig. 7.6)

1. Position: supine, with the arm abducted and the forearm semi-prone.
2. Incision: along the lateral border of the radius.
3. The superficial radial nerve is found in the superficial fascia accompanied by the cephalic vein.
4. Followed proximally, it can be seen entering the deep fascia and then disappears under the tendon of brachioradialis.
5. Distally, the superficial radial nerve enters the dorsum of the wrist and hand and branches into the terminal cutaneous branches.

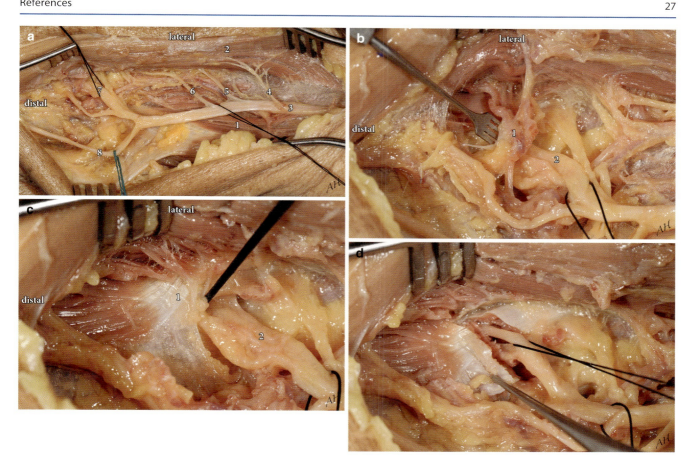

Fig. 7.5 Right radial nerve exposure in the proximal forearm. (**a**) The radial nerve (*3*) is found between the biceps brachii (*1*) medially and the brachioradialis (*2*) laterally. The two main terminal branches are the posterior interosseous nerve (PIN) (*7*) and the superficial radial nerve (*8*). Note the branches to the brachioradialis (*4*), extensor carpi radialis longus (ECRL) (*5*), and extensor carpi radialis brevis (ECRB) (*6*) take off before the origin of the PIN. (**b**) A vascular leash (leash of Henry) (*1*) can be compressive to the PIN (*2*) and should be ligated and cut. (**c**) The arcade of Fröhse (*1*) is the sharp aponeurotic edge of the supinator; it can also be compressive to the PIN (*2*). (**d**) PIN seen after full decompression

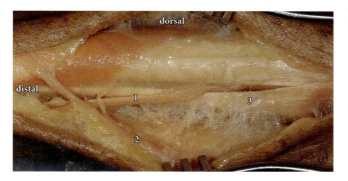

Fig. 7.6 Right superficial radial nerve (*1*) exposure in the distal forearm. *2* cephalic vein, *3* tendon of brachioradialis

References

1. Kinni V, Craig J, van Holsbeeck M, Ditmars D (2009) Entrapment of the posterior interosseous nerve at the arcade of Frohse with sonographic, magnetic resonance imaging, and intraoperative confirmation. J Ultrasound Med 28(6):807–812
2. Loizides A, Peer S, Ostermann S, Henninger B, Stampfer-Kountchev M, Gruber H (2011) Unusual functional compression of the deep branch of the radial nerve by a vascular branch (leash of Henry): ultrasonographic appearance. Rofo 183(2):163–166

Musculocutaneous Nerve

Surgical Anatomy

The musculocutaneous nerve (C5, C6, and C7) arises as one of the main branches of the lateral cord at the level of the coracoid process. It is lateral to the axillary artery. It supplies several branches to the coracobrachialis muscle before piercing it. More distally, it runs between the biceps brachii and the brachialis. It supplies the latter two muscles and then continues as the lateral antebrachial cutaneous nerve to supply the skin of the lateral aspect of the forearm. Various communications could occur between the median and musculocutaneous nerves in the arm [1].

Clinical Significance

Musculocutaneous nerve injury causes loss of elbow flexion. This typically occurs at the level of the lateral cord or the upper trunk. It could also be caused by avulsion of the C5 and C6 nerve roots. It is a common target for reinnervation (see Chap. 23).

Exposure

In the Axilla (Fig. 8.1)

1. Infraclavicular exposure as in Chap. 3.
2. The nerve is then followed laterally, which will require retracting or transecting the pectoralis major tendon.
3. It courses laterally to pierce the coracobrachialis muscle after giving off several branches to supply that muscle (Fig. 8.2).

Electronic supplementary material The online version of this chapter 10.1007/978-3-319-14520-4_8 contains supplementary material, which is available to authorized users.

Fig. 8.1 The left musculocutaneous nerve (*8*) is one of the two terminal branches of the lateral cord (*4*). It pierces the coracobrachialis after supplying it. The ulnar nerve (*10*) is one of the terminal branches of the medial cord (*5*). The median nerve (*9*) arises from one or more branches of the lateral and medial cords. *1* upper trunk, *2* middle trunk, *3* lower trunk, *6* posterior cord, *7* axillary artery

Fig. 8.2 The musculocutaneous nerve (*1*) as it enters the coracobrachialis (*2*). Note the comblike appearance of the branches to the coracobrachialis. Early branching can occur from higher location on the musculocutaneous nerve or the lateral cord. *3* median nerve, *4* axillary artery

In the Arm (Fig. 8.3)

1. Skin incision along the medial border of the biceps brachii muscle. One may try to palpate the brachial artery pulse and cut along it.
2. The incision is deepened through the superficial and deep fasciae.

3. The medial aspect of the biceps brachii is identified and retracted laterally; this uncovers the interval between the biceps brachii and brachialis muscles. The musculocutaneous nerve is found in that interval.
4. Followed proximally, we can identify the biceps brachii branch along the medial and deep surface of the muscle, and follow it until its origin from the musculocutaneous nerve.
5. Distally, the brachialis branch comes off and courses over the anterior surface of the muscle.
6. The main nerve continues laterally and distally as the lateral antebrachial cutaneous nerve. It pierces the deep fascia lateral to the tendon of the biceps brachii.

Reference

1. Kumar N, Guru A, D'Souza MR, Patil J, Nayak BS (2013) Incidences and clinical implications of communications between musculocutaneous nerve and median nerve in the arm - a cadaveric study. West Indian Med J 62(8):744–747

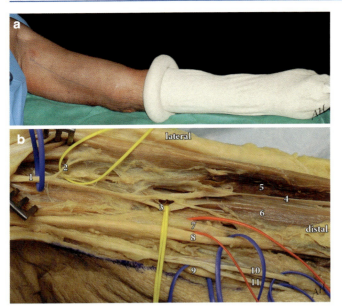

Fig. 8.3 (a) Musculocutaneous, median, and ulnar nerve exposure in the inner aspect of the left arm. The patient is positioned supine with the arm abducted. Incision is made along the medial aspect of the arm, with the neurovascular bundle. Pulse can be palpated to mark the incision. (b) In this cadaveric dissection, the musculocutaneous (*1*), median (*8*), and ulnar (*9*) nerves are exposed in the arm. The median and ulnar nerves have no major branches in the arm; the musculocutaneous nerve gives off a biceps brachii branch (*2*), a brachialis branch (*3*), and the lateral antebrachial cutaneous nerve (*4*). The musculocutaneous nerve courses between the biceps brachii (*5*) and brachialis (*6*) muscles. *7* brachial artery, *10* medial antebrachial cutaneous nerve, *11* medial brachial cutaneous nerve

Median Nerve

Surgical Anatomy

The median nerve (C5, C6, C7, C8, and T1) is formed by contributions from the medial and lateral cords in the axilla. It courses anterior to the axillary and then brachial arteries. It enters the cubital fossa anterior to the brachialis and deep to the bicipital aponeurosis (lacertus fibrosus). It gives off its first branches in the upper forearm. It supplies most of the flexor muscles of the forearm: pronator teres, flexor carpi radialis, palmaris longus, and flexor digitorum superficialis (sublimis). The anterior interosseous nerve (AIN) comes off the lateral aspect of the median nerve and supplies the flexor pollicis longus, lateral half of flexor digitorum profundus, and pronator quadratus. It has no cutaneous distribution. The median nerve courses between the two heads of pronator teres and then the two heads of flexor digitorum superficialis. It then enters the carpal tunnel with the flexor tendons under the flexor retinaculum (transverse carpal ligament). The palmar cutaneous branch typically arises proximal to the wrist and courses superficial to the flexor retinaculum. In the hand, the median nerve gives off a recurrent motor branch (palmar recurrent) that supplies the muscles of the thenar eminence: flexor pollicis brevis, abductor pollicis brevis, and opponens pollicis. It also supplies the lateral 2 lumbrical muscles and then divides into digital cutaneous branches that supply the palmar surface of the lateral three and half fingers and the dorsal surface of their distal phalanges.

Anatomical Variations

Martin-Gruber anastomosis: a connection from median to ulnar nerves in the forearm

Marinacci communication: ulnar to median nerves in the forearm

Riche-Cannieu anastomosis: deep ulnar to palmar recurrent branch of median nerve in the hand

Berrettini anastomosis: connection between ulnar and median digital nerves in the palm [1]

Clinical Significance

Carpal tunnel syndrome typically spares sensation in the palm since the palmar cutaneous nerve typically passes superficial to the transverse carpal ligament. Advanced cases can present with thenar muscle weakness and atrophy (Fig. 9.1).

AIN lesions cause inability to flex the interphalangeal joint (IPJ) of the thumb and the distal interphalangeal joint

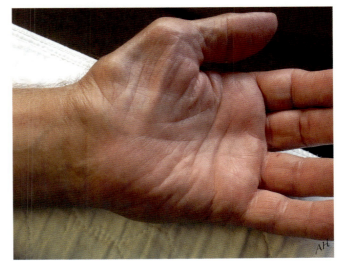

Fig. 9.1 Thenar atrophy from severe carpal tunnel syndrome. Such patients would also typically have decreased sensation in the lateral 3.5 fingers, sparing the palm

Electronic supplementary material The online version of this chapter 10.1007/978-3-319-14520-4_9 contains supplementary material, which is available to authorized users.

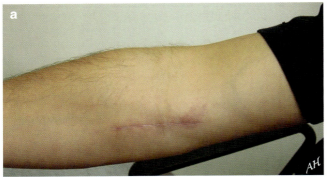
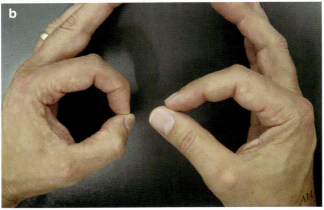

Fig. 9.2 Following resection of a right median nerve schwannoma (incision shown in **a**), the patient suffered from anterior interosseous nerve (AIN) palsy. Note the inability to make the "OK" sign with the right hand (**b**). There was no sensory loss

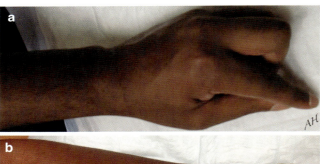
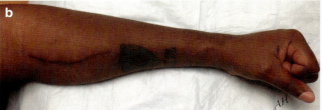

Fig. 9.3 (**a**) Benediction hand: after a stab wound to the axilla, the patient is unable to make a fist with the left hand. The lesion involved the median nerve at its origin from the lateral cord. (**b**) Six months after surgical repair: sural nerve grafting to the lateral cord and extensor carpi radialis brevis nerve transfer to the anterior interosseous nerve

(DIP) of the index finger resulting in inability to make the "OK" sign (Fig. 9.2). There is no sensory loss.

More proximal median nerve lesions cause inability to make a fist. The medial two fingers are still able to flex since the ulnar nerve is intact, resulting in the benediction hand (Fig. 9.3).

Struthers' ligament can arise from the medial aspect of the humerus above the medial epicondyle with or without a bony spur and may cause compression of the median nerve [2].

Exposures

In the Arm (See Fig. 8.3)

1. Skin incision along the medial border of the biceps brachii muscle. One may try to palpate the brachial artery pulse and cut along it.
2. The incision is deepened through the superficial and deep fasciae.
3. The median nerve is usually found superficial to the brachial artery.
4. Followed proximally towards the axilla under the pectoralis major tendon, it can be traced back to the contributions from the medial and lateral cords.
5. Distally, it can be followed between the biceps brachii and brachialis muscles.

In the Cubital Fossa (Fig. 9.4)

1. Midline anterior incision z-shaped to avoid crossing the elbow crease.
2. The antecubital vein may be encountered in the superficial fascia; it can be ligated and cut.
3. The deep fascia is opened and then the bicipital aponeurosis.
4. This exposes the contents of the cubital fossa with the brachial artery dividing into radial and ulnar arteries and the median nerve medial to the artery.
5. The median nerve gives off branches to the superficial flexor muscles except flexor carpi ulnaris.
6. The AIN comes off the radial aspect of the median nerve.
7. The median nerve and AIN leave the cubital fossa between the two heads of the pronator teres and then course under the tendinous origin of flexor digitorum superficialis (i.e., sublime bridge) [3].

Exposures

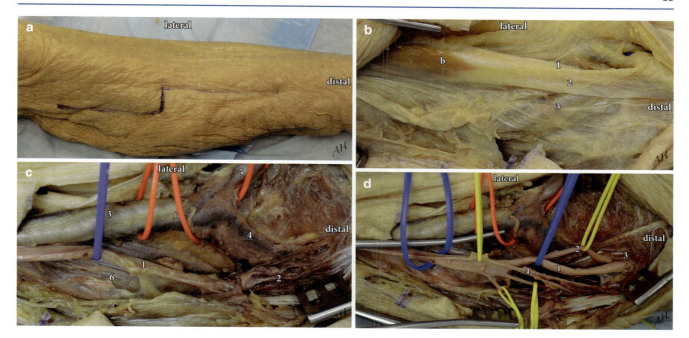

Fig. 9.4 Median nerve exposure in the left cubital fossa. (**a**) Skin incision. (**b**) *b* biceps brachii, *1* biceps tendon, *2* bicipital aponeurosis (lacertus fibrosus), *3* brachial artery. (**c**) The bicipital aponeurosis has been opened, exposing the median nerve (*1*) as it enters the pronator teres (*2*), the brachial artery (*3*), and its venae comitantes, as well as its bifurcation into ulnar (*4*) and radial (*5*) vessels. The brachialis (*6*) is seen in the floor of the cubital fossa. (**d**) Median nerve (*1*) branches are identified: *2* anterior interosseous nerve (AIN), the only branch arising from the radial aspect of the median nerve. Both median nerve and AIN travel between the 2 heads of flexor digitorum superficialis (*3*) (sublime bridge). Pronator teres branches (*4*) are the first ones to arise from the median nerve

In the Carpal Tunnel (Fig. 9.5)

1. Anterior midline incision from the distal wrist crease to the thenar crease along the radial aspect of the ring finger.
2. The palmar cutaneous branch should be protected. It may course superficial to or within the transverse carpal ligament.
3. The palmar aponeurosis is the extension of the palmaris longus tendon. It can be opened along the skin incision or retracted to one side.
4. The transverse carpal ligament is then sharply opened using a no. 15 blade scalpel.
5. This exposes the median nerve. A Woodson tool is then introduced under the ligament to protect the nerve while the remaining part of the ligament is being cut. Dissection should remain on the ulnar side of the nerve to protect the palmar recurrent branch.
6. The decompression is then extended proximally and distally using the Metzenbaum scissors.
7. Once the perineural fat is identified, this delineates the distal extent of the decompression to protect the superficial palmar arch.
8. The procedure could also be performed endoscopically.

In the Hand

1. Incisions in the hand and fingers are made in a zigzag fashion to avoid contractures.
2. The digital nerves course along the medial and lateral aspects of the fingers along with the digital vessels.

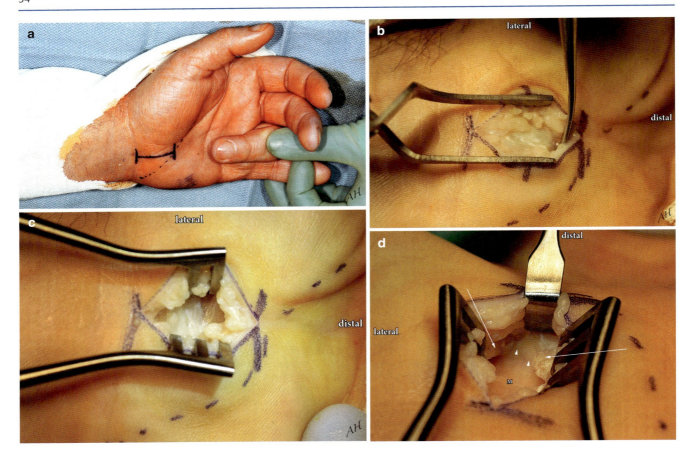

Fig. 9.5 Left carpal tunnel release. (**a**) Incision is made from the distal wrist crease till the Kaplan's cardinal line (from the thenar crease to the pisiform bone), in line with the radial aspect of the ring finger. (**b**) The palmar aponeurosis is exposed. (**c**) The transverse carpal ligament is now exposed after the palmar aponeurosis has been cut and retracted. Thenar muscle fibers are seen attaching laterally. (**d**) The hand is now rotated 90 degrees counterclockwise. The transverse carpal ligament (*arrows*) has been opened, thus decompressing the median nerve (*M*). Note that the fat (*arrowheads*) around the nerve distally marks the distal extent of the decompression. If a tourniquet is used, it is better to release it before closure, for adequate hemostasis

Complications of Carpal Tunnel Release

1. Failure to improve the symptoms from inadequate decompression
2. Recurrent symptoms from scar tissue
3. Pain from injury to the palmar cutaneous branch
4. Thumb paralysis from injury to the palmar recurrent branch
5. More rarely, injury to the median or ulnar nerves
6. Vascular injury to the superficial palmar arch
7. Wound infection or wound dehiscence especially in diabetic patients

References

1. Unver Dogan N, Uysal II, Karabulut AK, Seker M, Ziylan T (2010) Communications between the palmar digital branches of the median and ulnar nerves: a study in human fetuses and a review of the literature. Clin Anat 23(2):234–241
2. Bilecenoglu B, Uz A, Karalezli N (2005) Possible anatomic structures causing entrapment neuropathies of the median nerve review. Acta Orthop Belg 71(2):169–176
3. Tubbs RS, Marshall T, Loukas M, Shoja MM, Cohen-Gadol AA (2010) The sublime bridge: anatomy and implications in median nerve entrapment. J Neurosurg 113(1):110–112

Ulnar Nerve

Surgical Anatomy

The ulnar nerve (C7, C8, and T1) is one of the main branches of the medial cord of the brachial plexus in the axilla. It courses medial to the axillary artery and then posterior to the brachial artery. It enters the posterior compartment of the arm dorsal to the medial intermuscular septum and then the medial epicondyle. It enters the forearm between the two heads of flexor carpi ulnaris (when the ulnar head is present), to course between it, and the flexor digitorum profundus, medial to the ulnar artery. The first branches arise in the forearm supplying flexor carpi ulnaris and the medial half of flexor digitorum profundus. It enters Guyon's canal at the wrist, lateral to the pisiform bone. It gives off a superficial branch that supplies the palmaris brevis and the palmar aspect of the medial one and half fingers as well as the medial aspect of the palm and a deep motor branch that supplies the hypothenar muscles: abductor digiti minimi, flexor digiti minimi, and opponens digiti minimi. It also supplies the medial 2 lumbricals, the interossei, the adductor pollicis, as well as part of the flexor pollicis brevis. The dorsal cutaneous branch arises in the lower forearm to supply the dorsum of the medial one and half fingers as well as the medial aspect of the dorsum of the hand. Another palmar cutaneous branch may arise just proximal to Guyon's canal and supply the ulnar aspect of the palm proximally.

Clinical Significance

Ulnar nerve injuries at the wrist spare the dorsal sensation on the medial hand and the medial one and half fingers, flexor carpi ulnaris, and the medial half of flexor digitorum profundus. It results in ulnar claw hand (Fig. 10.1) with inability to extend the interphalangeal joints of the medial two fingers (lumbrical paralysis), hypothenar atrophy, and atrophy of the first dorsal interosseous space. Paradoxically with a more proximal lesion at the elbow, the clawing is less pronounced due to loss of flexor digitorum profundus function (ulnar nerve paradox). Wartenberg's sign (Fig. 10.2a) refers

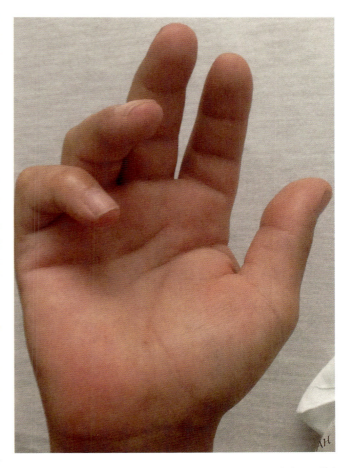

Fig. 10.1 Ulnar claw hand: the patient is unable to extend the medial 2 fingers from lumbrical paralysis

Electronic supplementary material The online version of this chapter 10.1007/978-3-319-14520-4_10 contains supplementary material, which is available to authorized users.

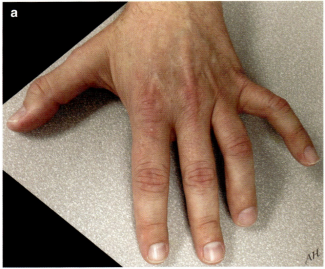

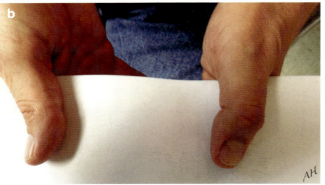

Fig. 10.2 (**a**) Wartenberg's sign. The patient is unable to adduct the little finger from paralysis of the interossei. Note also atrophy of the first dorsal interosseous and the hypothenar muscles. All are signs of ulnar neuropathy. (**b**) Positive Froment's sign in the left hand. The patient is unable to adduct the thumb from ulnar neuropathy; he is flexing the interphalangeal joint of the thumb instead (intact flexor pollicis longus supplied by AIN)

to the persistent abduction of the little finger by the extensor digiti minimi (quinti) (radial nerve) in the absence of the palmar interossei function. Froment's sign (Fig. 10.2b) refers to the flexion of the interphalangeal joint of the thumb (flexor pollicis longus supplied by the anterior interosseous nerve) to compensate for weakness of the adductor pollicis, when the patient tries to hold a flat object between the thumb and index fingers. Anconeus epitrochlearis is an anomalous accessory muscle that attaches to the medial epicondyle and olecranon process. It can cause ulnar nerve compression at the elbow [1]. The arcade of Struthers is a fibrous band between the medial head of triceps and medial intermuscular septum. It can cause persistent or recurrent symptoms after ulnar nerve surgery. Its presence is controversial. Osborne's band, fascia between the two heads of flexor carpi ulnaris, is another potential site of entrapment for the ulnar nerve. Ulnar subluxation occurs when the ulnar nerve dislocates anteriorly around the medial epicondyle with elbow flexion. Triceps snapping occurs when the medial head of triceps dislocates (snaps) around the medial epicondyle with elbow flexion. For ulnar nerve entrapment at the elbow, most randomized trials didn't show benefit for transposition over simple decompression [2]. Neurogenic thoracic outlet syndrome is an important differential diagnosis for ulnar neuropathy. It can manifest with Gilliatt-Sumner hand with atrophy of the thenar, hypothenar, and interossei muscles (Fig. 10.3).

Exposures

In the Arm (See Fig. 8.3)

1. Skin incision along the medial border of the biceps brachii muscle. One may try to palpate the brachial artery pulse and cut along it.
2. The incision is deepened through the superficial and deep fasciae.
3. The ulnar nerve is usually found posterior to the brachial artery.
4. Followed proximally towards the axilla under the pectoralis major tendon, we can trace it to its origin from the medial cord.
5. Distally, it enters the posterior arm behind the medial intermuscular septum and then travels posterior to the medial epicondyle.

Around the Elbow

1. Position: supine with the arm abducted and externally rotated and the elbow flexed.
2. Incision is made between the medial epicondyle and the olecranon process. It is extended cranially and caudally along the course of the nerve. A short incision is used for a simple decompression, but a longer (20 cm) incision is used for a transposition.
3. The superficial fascia is opened. Care is taken not to injure the medial antebrachial cutaneous nerve.
4. Before opening the deep fascia, one should palpate the ulnar nerve against the medial epicondyle. It is important to open the deep fascia right on top of the nerve to avoid missing the nerve. Using forceps to lift up the fascia and tenotomy scissors, the incision in the deep fascia is deepened until the nerve sheath is opened and the nerve is exposed.
5. Once in the correct plane, the nerve can be unroofed proximally and distally. Ulnar nerve branches start below the medial epicondyle. To protect these, dissection should be started on the surface of the nerve and not on its sides.

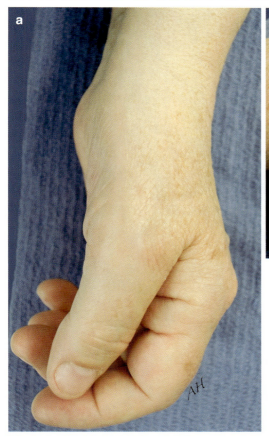
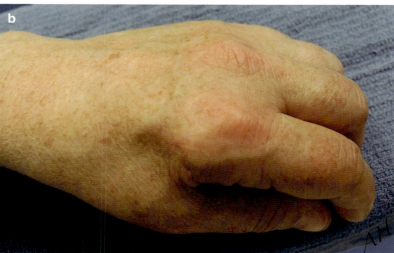

Fig. 10.3 Gilliatt-Sumner hand with atrophy of the thenar eminence (**a**) and interossei (**b**). These are supplied by the median and ulnar nerves, respectively. This localizes the lesion to the medial cord or the lower trunk of the brachial plexus; the latter can be seen in neurogenic thoracic outlet syndrome

6. Distally, the fascia overlying the flexor carpi ulnaris is opened, the two heads of the muscle separated, and the fascia underneath it opened (Fig. 10.4).
7. For a transposition (Fig. 10.5), the medial intermuscular septum should be resected and the ulnar branches dissected off the nerve in an interfascicular fashion, to avoid tension, and then the nerve is transposed subcutaneously or submuscularly.

At Guyon's Canal (Fig. 10.6)

1. Position: supine, with the arm abducted and supinated.
2. Incision is made lateral to the flexor carpi ulnaris tendon and pisiform, crossing the wrist obliquely, and then along the hypothenar eminence.
3. The flexor carpi ulnaris tendon is identified at the proximal part of the incision and retracted medially. The ulnar nerve with the artery lateral to it enters Guyon's canal lateral to the pisiform bone.
4. Guyon's canal is then unroofed and the nerve is followed until it branches.
5. The deep motor branch is followed under the hypothenar muscle origin, which is released for full decompression.

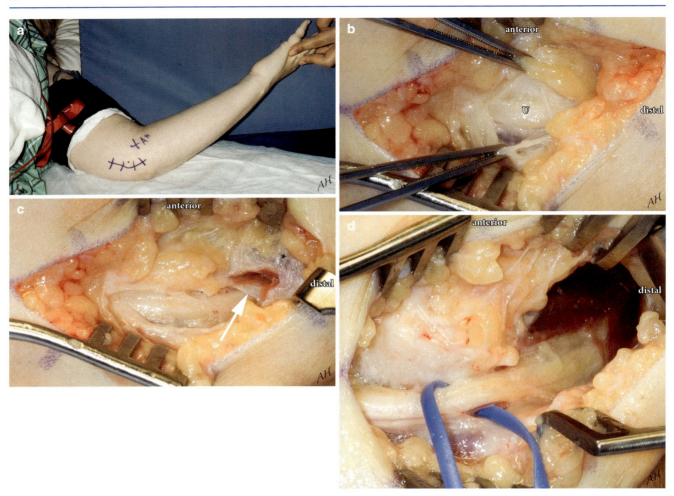

Fig. 10.4 Ulnar decompression at the elbow. (**a**) The patient is positioned supine with the arm abducted and externally rotated and the elbow flexed. Incision is made between the medial epicondyle and the olecranon. Tourniquet use is optional. (**b**) The nerve sheath is opened, and the ulnar nerve (*U*) is exposed. (**c**) The fascia (*) superficial to flexor carpi ulnaris is opened, the 2 heads are separated from each other, and the fascia (*arrow*) deep to the muscle is also opened. (**d**) The ulnar nerve (*blue loop*) is completely unroofed proximally and distally. There is no need for a 360° decompression; this may cause iatrogenic subluxation

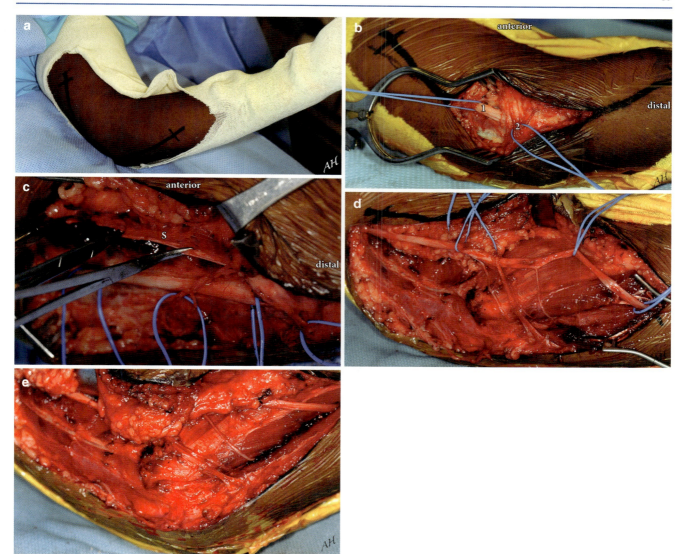

Fig. 10.5 Subcutaneous transposition of the ulnar nerve. (**a**) Position is the same as for a decompression; incision is longer, 10 cm on either side of the medial epicondyle. (**b**) The ulnar nerve (*1*) is exposed as above; the medial antebrachial cutaneous nerve (*2*) is identified and protected. (**c**) The medial intermuscular septum (*S*) is dissected free and resected. (**d**) Interfascicular dissection of the ulnar nerve branches increases the length of these branches and allows transposition without tension. Note the subcutaneous pocket created between the subcutaneous fat and the fascia overlying the flexor muscles. (**e**) Once the nerve is transposed, it is kept in place by suturing the fat to the fascia

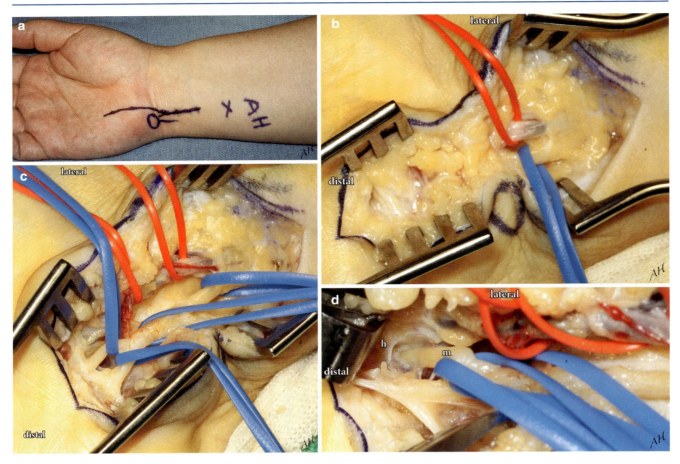

Fig. 10.6 Right Guyon's canal decompression. (**a**) The patient is positioned supine with the arm abducted and supinated on a hand table. The incision is lateral to the flexor carpi ulnaris (FCU) tendon and pisiform bone and extended into the palm lateral to the hypothenar eminence. (**b**) The FCU tendon is retracted medially; the ulnar nerve (*blue loop*) and vessels (*red loop*) are then exposed. These are seen entering the Guyon's canal distally. (**c**) After opening the Guyon's canal, the ulnar nerve and vessels are seen decompressed. (**d**) The deep motor branch (*m*) is followed until it enters under the hypothenar arch (*h*), which needs to be decompressed as well

In the Hand

1. Incisions in the hand and fingers are made in a zigzag fashion to avoid contractures.
2. The digital nerves course along the medial and lateral aspects of the fingers with the digital vessels.

References

1. Nellans K, Galdi B, Kim HM, Levine WN (2014) Ulnar neuropathy as a result of anconeus epitrochlearis. Orthopedics 37(8):e743–745
2. Bartels RH, Verhagen WI, van der Wilt GJ, Meulstee J, van Rossum LG, Grotenhuis JA (2005) Prospective randomized controlled study comparing simple decompression versus anterior subcutaneous transposition. Neurosurgery 56(3):522–530

Suprascapular Nerve

Surgical Anatomy

The suprascapular nerve arises from the upper trunk of the brachial plexus (C5 and C6). The upper trunk division almost looks like a trifurcation with the suprascapular nerve, posterior division, and then anterior division arranged from cranial to caudal, respectively. It travels posteriorly towards the suprascapular notch. It passes underneath the suprascapular ligament, which separates the nerve from the suprascapular artery. Occasionally, the suprascapular ligament is completely ossified transforming the notch into a foramen [1, 2]. The nerve then travels deep to the supraspinatus, which it supplies, and then through the spinoglenoid notch to the undersurface of the infraspinatus, which it also supplies.

Clinical Significance

The suprascapular nerve can be damaged by entrapment at the suprascapular notch, which affects the supraspinatus and infraspinatus muscles causing weak shoulder abduction and external rotation and muscle atrophy (Fig. 11.1). More distal compression at the spinoglenoid notch would involve the infraspinatus only. It can also be damaged in upper trunk stretch or avulsion injuries, which would also affect the deltoid, biceps brachii, and brachialis muscles.

Exposure

Anterior

See Chap. 2. The suprascapular nerve branches off distally from the cranial aspect of the upper trunk of the brachial plexus.

Electronic supplementary material The online version of this chapter 10.1007/978-3-319-14520-4_11 contains supplementary material, which is available to authorized users.

Posterior

1. The patient is positioned laterally on a beanbag (Fig. 11.2).
2. Skin incision: along the upper border of the spine of the scapula.
3. The incision is deepened through the superficial and deep fasciae.
4. The trapezius is opened horizontally along its fibers, above the spine of the scapula (Fig. 11.3). A self-retaining retractor is placed.
5. Another layer of deep fascia is encountered; once opened, this exposes the supraspinatus muscle.
6. The supraspinatus muscle is retracted caudally; this exposes the upper border of the scapula. Followed laterally, one can feel the suprascapular notch.
7. A Doppler ultrasound probe can be used to identify the suprascapular artery. The artery is retracted to visualize the suprascapular ligament (Fig. 11.4).
8. The ligament is opened sharply with scissors or if it is ossified with Kerrison rongeurs. The notch itself

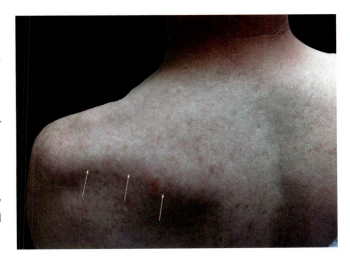

Fig. 11.1 Left suprascapular nerve entrapment at the suprascapular notch. Note the prominent spine of the left scapula (*arrows*) due to atrophy of supraspinatus and infraspinatus muscles

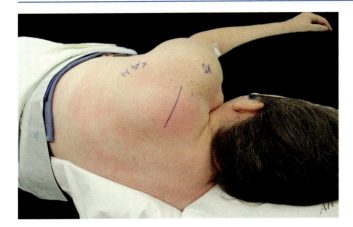

Fig. 11.2 A patient positioned lateral for left suprascapular decompression. Prone position could also be used

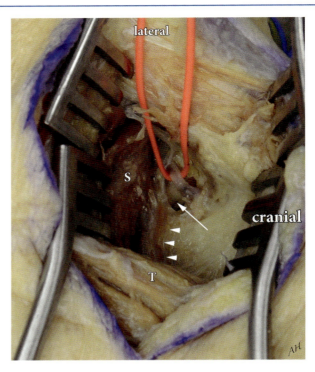

Fig. 11.4 The supraspinatus (*S*) is retracted caudally, exposing the upper border of the scapula (*arrowheads*). The suprascapular artery (*red loop*) is observed superficial to the transverse scapular ligament (*arrow*) and suprascapular notch. *T* trapezius

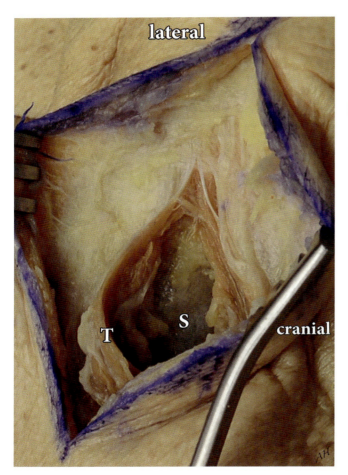

Fig. 11.3 Cadaveric dissection of the left suprascapular nerve. The trapezius (*T*) is opened along its fibers. This exposes the fascia covering the supraspinatus (*S*)

can be enlarged with Kerrison rongeurs for full decompression.
9. The suprascapular nerve can then be identified and followed proximally and distally (Fig. 11.5).
10. At the spinoglenoid notch, the supraspinatus muscle is retracted cranially to expose the nerve as it enters the spinoglenoid notch (Fig. 11.6). More inferior exposure would require separation of the infraspinatus muscle from the undersurface of the spine of the scapula. The nerve can be identified as it passes from the spinoglenoid notch to the undersurface of the infraspinatus muscle (Fig. 11.7). Using Kerrison rongeurs, the spinoglenoid notch can be enlarged for further decompression.

Exposure

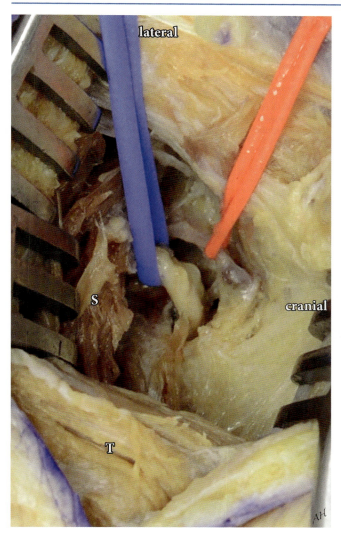

Fig. 11.5 The suprascapular ligament has been opened. The suprascapular nerve (*blue loop*) is now decompressed. *S* supraspinatus, *T* trapezius

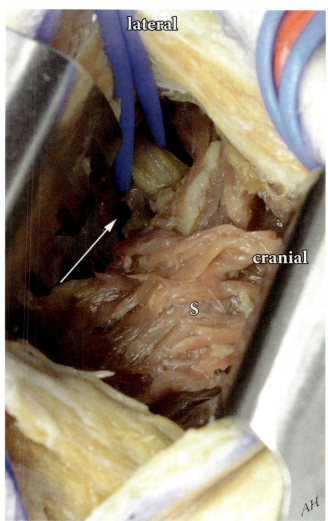

Fig. 11.6 If further decompression is needed, the supraspinatus (*S*) is retracted cranially exposing the suprascapular nerve (*blue loop*) in the spinoglenoid notch (*arrow*)

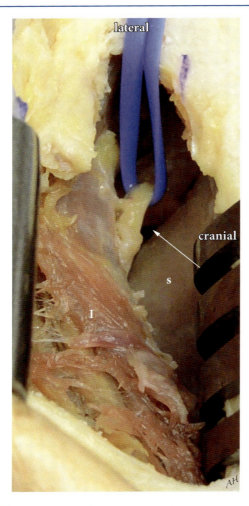

Fig. 11.7 More caudally, the infraspinatus (*I*) can be separated from the spine of the scapula (*s*), exposing the suprascapular nerve (*blue loop*) coming out of the spinoglenoid notch (*arrow*)

References

1. Polguj M, Sibinski M, Grzegorzewski A, Waszczykowski M, Majos A, Topol M (2014) Morphological and radiological study of ossified superior transverse scapular ligament as potential risk factor of suprascapular nerve entrapment. Biomed Res Int 2014:613601
2. Kumar A, Sharm A, Singh P (2014) Anatomical study of the suprascapular notch: quantitative analysis and clinical considerations for suprascapular nerve entrapment. Singapore Med J 55(1):41–44

Spinal Accessory Nerve

Surgical Anatomy

Although the accessory nerve is the 11th cranial nerve, the spinal component will be discussed here due to its importance in nerve injuries and repair. The spinal accessory nerve arises from the upper cervical spinal cord (C1–C5), ascends between the dentate ligament and the posterior cervical rootlets, then through the foramen magnum joins the cranial part, and exits through the jugular foramen. It runs through or deep to the sternocleidomastoid (SCM) muscle. It courses subcutaneously in the posterior triangle of the neck from the posterior border of the SCM cranially to the anterior border of the trapezius muscle caudally. It supplies both the SCM and the trapezius muscles.

Clinical Significance

The spinal accessory nerve is usually intact with brachial plexus injuries. It can be used as a donor (distal to some trapezius branches) for nerve transfers to the suprascapular nerve or a free muscle flap.

The nerve can be injured in the posterior triangle of the neck (Fig. 12.1) during lymph node biopsies or lipoma resections, causing shoulder pain and paralysis of the trapezius muscle. This can result in scapular winging.

Exposure

During Brachial Plexus Exposure (Fig. 12.2)

1. Supraclavicular brachial plexus exposure as in Chap. 2.
2. The nerve is found within the fat pad along the anterior border of the trapezius muscle.
3. Dissection is carried along the undersurface of the trapezius using Metzenbaum scissors. Electrical stimulation is used to identify the nerve. Of note, some cutaneous cervical plexus branches may take a similar course as the spinal accessory, but they are usually superficial to the trapezius and are negative for stimulation.
4. If used as a donor, the spinal accessory nerve should be harvested distal to some trapezius branches to avoid postoperative weakness.

Spinal Accessory Alone Exposure

1. Position: supine with the head turned to the contralateral side.
2. The nerve trajectory in the posterior triangle of the neck should be drawn first. This is represented by a line joining the posterior border of the SCM about 5 cm below the mastoid process and the anterior border of the trapezius about 5 cm above the midpoint of the clavicle (Fig. 12.3) [1].
3. Incision is made along this line through the skin and superficial fascia. If there is a previous incision from another procedure that damaged the nerve, it should be incorporated into the new incision. The nerve is very superficial and the use of electrocautery is discouraged to avoid accidental damage to the nerve.
4. The nerve is found in the superficial fascia as it courses from ventral and cranial to dorsal and caudal. At the posterior border of the SCM, the nerve is about 1.5 cm cranial to the great auricular nerve. The great auricular nerve winds around the posterior border of the SCM from deep to superficial and ascends vertically towards the angle of the mandible and ear.
5. The spinal accessory nerve can be followed anteriorly until the undersurface of the SCM and posteriorly until the undersurface of the trapezius muscle.

Electronic supplementary material The online version of this chapter 10.1007/978-3-319-14520-4_12 contains supplementary material, which is available to authorized users.

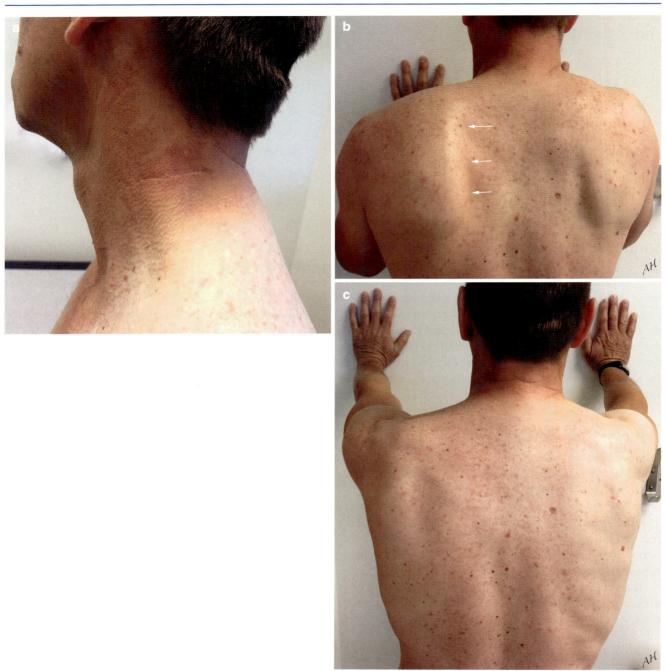

Fig. 12.1 (**a**) Incision for resection of a subcutaneous lipoma of the neck. (**b**) This resulted in partial injury of the spinal accessory nerve and left scapular winging (arrows). Note the winging is more pronounced with elbow flexion, rather than with elbow extension (**c**)

Reference

1. Tubbs RS, Salter EG, Wellons JC III, Blount JP, Oakes WJ (2005) Superficial landmarks for the spinal accessory nerve within the posterior cervical triangle. J Neurosurg Spine 3:375–378

Fig. 12.2 Left spinal accessory nerve (*1*) exposure during a brachial plexus exposure. The nerve is found within the fat (*F*) anterior to the trapezius muscle (*3*). *2* sternocleidomastoid, *4* phrenic nerve, *5* upper trunk of brachial plexus, *6* nerve to the subclavius, *7* suprascapular nerve, *8* lateral cord of brachial plexus, *9* musculocutaneous nerve, *10* median nerve, *11* lateral pectoral nerve. Note this is arising from the anterior division of the upper trunk rather than the lateral cord. *12* upper and lower subscapular nerves, *13* axillary nerve, *14* radial nerve. Branches of the cervical plexus (*15*) may mimic the spinal accessory nerve. The spinal accessory is deep to the trapezius and has a wavy course

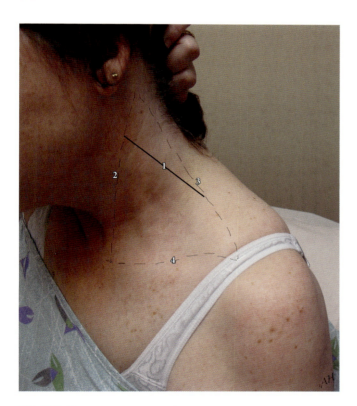

Fig. 12.3 Trajectory of the spinal accessory nerve (*1*) in the posterior triangle of the neck. It travels from 5 cm below the mastoid process along the posterior border of sternocleidomastoid (*2*) to the anterior border of trapezius (*3*), at 5 cm above the midclavicular point. *4* clavicle

Long Thoracic and Thoracodorsal Nerves

Surgical Anatomy

The long thoracic nerve arises from the posterior aspect of C5, C6, and often C7 ventral rami. It travels within the scalenus medius muscle. It then runs over the lateral surface of the serratus anterior supplying it with multiple branches.

The thoracodorsal nerve (C6, C7, and C8) arises from the posterior cord. It courses anterior to the latissimus dorsi muscle along with the corresponding vessels and supplies the muscle.

Clinical Significance

Injury to the long thoracic nerve can occur during mastectomies. The nerve is also usually involved in brachial plexitis (Parsonage-Turner syndrome). This causes scapular winging especially with the arms in full extension (Fig. 13.1).

The long thoracic and thoracodorsal nerves course close enough to each other that one of them can be used to neurotize the other, usually posterior division of the thoracodorsal nerve to long thoracic nerve [1].

Exposure

1. Position: supine with the arm abducted at about 90°.
2. Incision in the midaxillary line (Fig. 13.2).
3. The superficial fascia is opened.
4. The deep fascia overlying the serratus anterior is opened.

Electronic supplementary material The online version of this chapter 10.1007/978-3-319-14520-4_13 contains supplementary material, which is available to authorized users.

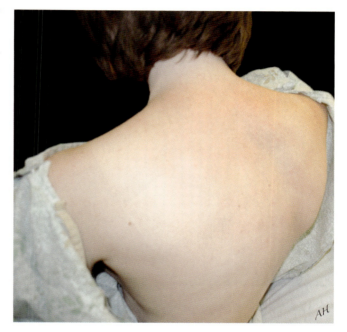

Fig. 13.1 Right scapular winging from long thoracic nerve damage

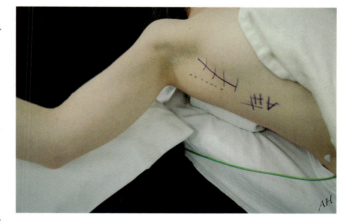

Fig. 13.2 Position and incision for exposure of the long thoracic and thoracodorsal nerves

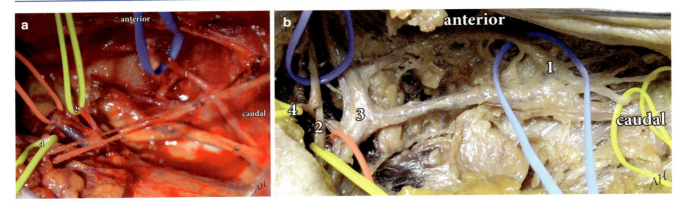

Fig. 13.3 The long thoracic nerve (*1*) is found anteriorly deep to the fascia covering the serratus anterior. The thoracodorsal nerve has two branches: anterior (*2*) and posterior (*4*) which is found along the thoracodorsal vessels (*3*) towards the posterior axillary fold. (**a**) Surgical view. (**b**) Cadaveric dissection

5. The long thoracic nerve can be seen on the surface of the serratus anterior muscle, sequentially giving off branches to supply different digitations of the muscle (Fig. 13.3).
6. To expose the thoracodorsal nerve, the deep fascia overlying the latissimus dorsi muscle is opened. A micro-Doppler can be used to locate the corresponding vessels that usually travel along the nerve.
7. Once identified, the nerve can be followed proximally and distally until its anterior and posterior divisions.
8. The posterior division of the thoracodorsal nerve can be used to reinnervate the long thoracic nerve (Fig. 13.4)

Reference

1. Ray WZ, Pet MA, Nicoson MC, Yee A, Kahn LC, Mackinnon SE (2011) Two-level motor nerve transfer for the treatment of long thoracic nerve palsy. J Neurosurg 115:858–864

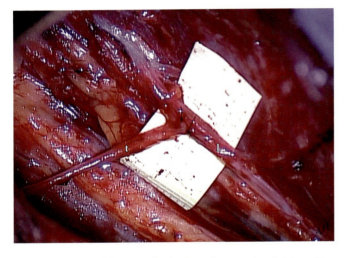

Fig. 13.4 End-to-side neurorrhaphy from the posterior division of the thoracodorsal nerve (*t*) to the long thoracic nerve (*l*)

Intercostal Nerves

Surgical Anatomy

The intercostal nerves arise from the ventral rami of the corresponding thoracic nerve roots. They run on the undersurface of the lower border of the corresponding rib in a neurovascular bundle with the following arrangement from cranial to caudal: vein, artery, and nerve (VAN).

They typically follow the course of the rib from dorsal to ventral. They supply the intercostal muscles and give off two cutaneous branches, one laterally and one anteriorly. They also send a collateral branch that follows the upper border of the next rib. Exceptions to this are the first and second thoracic ventral rami: T1 main contribution joins the brachial plexus lower trunk, a smaller branch continues as the first intercostal nerve. T2 lateral cutaneous branch forms the intercostobrachial nerve that supplies the skin of the floor of the axilla and medial upper arm.

Clinical Significance

The intercostal nerves are usually used to reinnervate the brachial plexus. They are good donors for nerve transfers. They can be used for the musculocutaneous nerve [1], axillary nerve, or to a free gracilis muscle flap. The cutaneous component can be used to graft to the median nerve in an attempt to regain sensation in the hand.

Exposure

1. Incision along the midaxillary line curving anteriorly along the ribs at about the sixth intercostal space (Fig. 14.1).
2. The incision is deepened through the superficial and deep fasciae.
3. A flap is raised separating the pectoralis major from the chest wall.
4. The ribs are identified. Typically, intercostal nerves 3–6 are the usual donors (Fig. 14.2).
5. The periosteum is opened sharply with a scalpel blade. A periosteal elevator is used to achieve a subperiosteal dissection of the anterior surface and upper and lower borders of the rib (Fig. 14.3).
6. A Doyen rib dissector is used to strip the periosteum from the undersurface of the rib. The pleura is very thin and

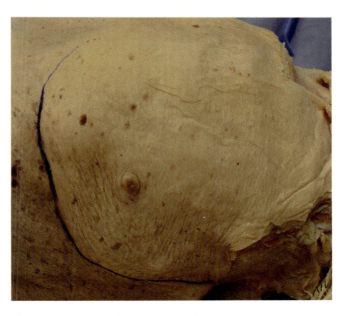

Fig. 14.1 Incision for intercostal nerve harvesting

Electronic supplementary material The online version of this chapter 10.1007/978-3-319-14520-4_14 contains supplementary material, which is available to authorized users.

Fig. 14.2 Once pectoralis major and minor are separated from the chest wall, the ribs can be exposed. *1* serratus anterior, *2* pectoralis minor, *3* pectoralis major

Fig. 14.3 A periosteal elevator is used to dissect the periosteum from the rib

Fig. 14.4 A Ray-Tec sponge is used to retract the rib cranially to expose the neurovascular bundle

Fig. 14.5 The intercostal nerve (*yellow loop*) has been dissected free. A nerve stimulator can be used to find the motor nerve

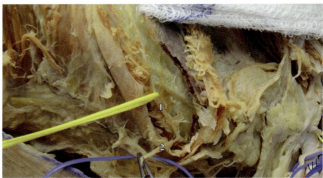

Fig. 14.6 Both the motor (*1*) and sensory (*2*) branches have been dissected

care should be taken not to violate the pleura. A Ray-Tec sponge is then used to retract the rib (Fig. 14.4).
7. The neurovascular bundle can then be identified at the lower border of the rib. The deep periosteum overlying the neurovascular bundle is opened sharply. Electrical stimulation can be used to find the nerve.
8. The nerve is then followed medially and laterally to gain as much length as possible (Fig. 14.5).
9. The lateral cutaneous branch takes off laterally almost perpendicular to the main nerve and can just be pulled through the superficial tissues (Fig. 14.6).

Complications

Pneumothorax: If the pleura is violated, it can be closed with a 3-0 Vicryl stitch over a small feeding tube. The tube can be pulled out during a Valsalva maneuver. Postoperative chest X-ray should be obtained. Larger pneumothorax will warrant chest tube placement.

Reference

1. Xiao C, Lao J, Wang T, Zhao X, Liu J, Gu Y (2014) Intercostal nerve transfer to neurotize the musculocutaneous nerve after traumatic brachial plexus avulsion: a comparison of two, three, and four nerve transfers. J Reconstr Microsurg 30(5):297–304

Part III
Lower Body Spinal Nerves

Lumbosacral Plexus

Surgical Anatomy

The lumbar plexus arises from the ventral rami of L1–L5 nerve roots. It forms within the psoas major muscle. The nerves are divided into three groups based on their relationship to the psoas muscle as they enter the retroperitoneal space.

Anterior to the Psoas Major Muscle

Genitofemoral nerve (L1, L2)

Lateral to the Psoas Major Muscle, from Cranial to Caudal

Iliohypogastric (T12, L1)
Ilioinguinal (L1)
Lateral femoral cutaneous nerve (L2, L3, posterior division)
Femoral nerve (L2–L4, posterior division)

Medial to the Psoas Major Muscle, from Cranial to Caudal

Obturator nerve (L2–L4, anterior division)
Accessory obturator nerve (seen in ~10 % of cases) (L2–L4, anterior division)
Lumbosacral trunk (L4, L5)

The sacral plexus arises from the ventral rami of L4–S4. It forms in the pelvis anterior to the piriformis muscle. The nerves are divided into four groups depending on their destination:
- *Buttock*: they exit the pelvis through the greater sciatic foramen:
  - Superior gluteal nerve (L4, L5, S1 above the piriformis to the gluteus medius, gluteus minimus, and tensor fasciae latae)
  - Inferior gluteal nerve (L5, S1, S2 below the piriformis to the gluteus maximus)
- *Lower extremity*: they exit the pelvis through the greater sciatic foramen below the piriformis muscle:
  - Sciatic nerve (tibial nerve, anterior division L4–S3; common peroneal nerve [common fibular nerve], posterior division L4–S2)
  - Posterior cutaneous nerve of the thigh (S1–S3)
- *Perineum*: it exits the pelvis through the greater sciatic foramen and enters the perineum through the lesser sciatic foramen:
  - Pudendal nerve (S2–S4)
- *Local*: three groups:
  - *Muscular*:
    - Piriformis
    - Obturator internus and superior gemellus
    - Quadratus femoris and inferior gemellus
  - Cutaneous:
    - Perforating cutaneous branches
    - Perineal branch of fourth sacral
  - *Parasympathetic*:
    - Pelvic splanchnic nerves (S2–S4 nervi erigentes)

Clinical Significance

The need to expose the lumbosacral plexus is much less common than the brachial plexus. Common pathologies include tumors rather than trauma. The approach usually involves an access surgeon (general, trauma, vascular, or colorectal surgeon). Occasionally, proximal lesions at a root level could be approached posteriorly via a paraspinal (Wiltse) approach,

---

Electronic supplementary material The online version of this chapter 10.1007/978-3-319-14520-4_15 contains supplementary material, which is available to authorized users.

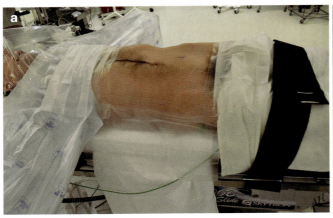 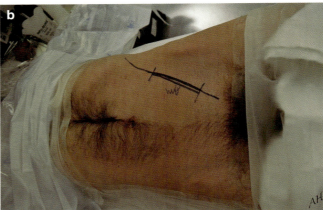

Fig. 15.1 Left lumbosacral plexus exposure. (**a**) The patient is positioned supine. (**b**) Oblique incision is made parallel to the external oblique muscle. The size and level of the incision depends on the pathology and location to be addressed. Large incisions can be muscle cutting. Small incisions should be muscle splitting

with various degrees of facetectomies depending on the intraspinal extent of the lesion, which may warrant spinal instrumentation.

Iatrogenic injuries to the lumbosacral plexus can occur during hip surgery (sciatic nerve, femoral nerve), pelvic surgery, or direct lateral interbody fusion (DLIF) for the lumbar spine (femoral nerve).

Perineurial spread of pelvic cancer could cause involvement of the lumbosacral nerve roots that can mimic sciatic pain [1].

Piriformis syndrome can cause sciatic pain from compression of the sciatic nerve at the piriformis [2].

Surgical Approaches

Retroperitoneal [3]

1. An access surgeon is usually required. The procedure can be done open or laparoscopic. Here we describe the open approach.
2. Position: supine (Fig. 15.1) or lateral on a beanbag.
3. Incision through the skin and superficial fascia: superficial fatty (Camper's) and deep membranous (Scarpa's).
4. Incision through the abdominal wall muscles—external oblique, internal oblique, and transversus abdominis—can be muscle splitting (Fig. 15.2) or cutting depending on the size on the incision.
5. Once the transversalis fascia is opened, the glistening yellow extraperitoneal fat is recognized. Further dissection in this plane will expose the quadratus lumborum and iliacus laterally and the psoas major muscle medially.
6. Nerves lateral to the psoas muscle from cranial to caudal include iliohypogastric, ilioinguinal, lateral femoral cutaneous, and femoral nerves. The fascia overlying the posterior abdominal wall muscles needs to be opened to expose the nerves. In addition, for the femoral nerve, the psoas muscle needs to be separated from the iliacus muscle.

Fig. 15.2 Muscle splitting approach shown in a cadaveric dissection. *E* external oblique, *I* internal oblique, *T* transversus abdominis, *t* transversalis fascia, *F* extraperitoneal fat, *i* ilioinguinal nerve

7. Anterior to the psoas major muscle is the genitofemoral nerve, which divides into its femoral and genital components.
8. Medial to the psoas muscle are the obturator nerve and the lumbosacral trunk. Their exposure will likely require mobilization of the iliac vessels (Fig. 15.3).
9. The sacral plexus is deeply located in the pelvis anterior to the piriformis muscle. Exposure would require mobilization of the colon and rectum in addition to the iliac vessels.

Posterior Paraspinal Approach (Wiltse) [4]

1. Position: prone.
2. X-ray localization is usually required prior to skin incision.

References

7. Electrical stimulation can be used to identify the nerve roots (electromyogram = EMG)
8. A partial or complete facetectomy can be performed as needed.

Complications

Retroperitoneal Approach

Vascular injury, ureteric injury, or bowel injury could occur. Incisional hernia could complicate, especially in muscle cutting approaches.

Wiltse Approach

Neuropathic pain from manipulating the dorsal root ganglion (DRG)

References

1. Hebert-Blouin MN, Amrami KK, Myers RP, Hanna AS, Spinner RJ (2010) Adenocarcinoma of the prostate involving the lumbosacral plexus: MRI evidence to support direct perineural spread. Acta Neurochir (Wien) 152(9):1567–1576
2. Filler AG (2008) Piriformis and related entrapment syndromes: diagnosis and management. Neurosurg Clin N Am 19(4):609–622
3. McAfee PC, Bohlman HH, Yuan HA (1985) Anterior decompression of traumatic thoracolumbar fractures with incomplete neurological deficit using a retroperitoneal approach. J Bone Joint Surg Am 67-A(1):89–104
4. Wiltse LL, Spencer CW (1988) New uses and refinements of the paraspinal approach to the lumbar spine. Spine 13(6):696–706

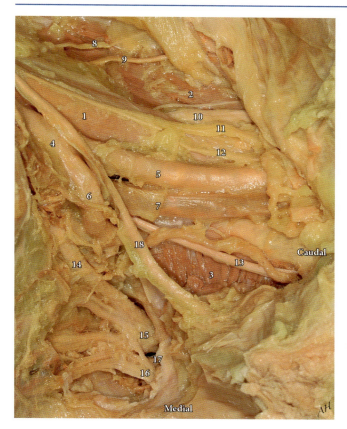

Fig. 15.3 Retroperitoneal view of the left lumbosacral plexus in a cadaver. *1* psoas major, *2* iliacus, *3* obturator internus, *4* common iliac artery, *5* external iliac artery, *6* internal iliac artery, *7* external iliac vein, *8* ilioinguinal nerve, *9* lateral femoral cutaneous nerve, *10* femoral nerve. Note that 8–10 exit at the lateral border of the psoas muscle. *11* femoral branch of the genitofemoral nerve, *12* genital branch of the genitofemoral nerve. Note that 11 and 12 are anterior to the psoas muscle. *13* obturator nerve, *14* lumbosacral trunk. Note that 13 and 14 are medial to the psoas muscle. *15* sciatic nerve, *16* pudendal nerve, *17* posterior cutaneous nerve of the thigh, *18* ureter

3. Incision about 4 cm lateral to the midline.
4. Incision is deepened through the superficial and deep fasciae.
5. The paraspinal muscles are opened to expose the transverse processes of the corresponding vertebrae. A self-retaining retractor is placed.
6. The inter-transverse membrane is opened.

Femoral Nerve

Surgical Anatomy

The femoral nerve arises from the posterior division of the ventral rami of L2–L4. It forms within the psoas muscle and exits lateral to the muscle supplying it and the iliacus muscle. It courses in the retroperitoneum and then enters the thigh below the inguinal ligament with the corresponding vessels. The vein (V) is most medial, then the artery (A), and then the nerve (N) (VAN), respectively. The femoral nerve proper has a short course. It divides into multiple muscular branches to the quadriceps femoris and sartorius, medial and intermediate cutaneous nerves of the thigh, as well as the saphenous nerve. The saphenous nerve courses along the femoral artery in the adductor canal and then along the long saphenous vein on the medial aspect of the leg.

Clinical Significance

The saphenous nerve, similar to the sural and the superficial radial nerves, can be used for nerve biopsies (6 cm) or as a donor for nerve grafts. Neuromas of the infrapatellar branch of the saphenous nerve are a common cause of pain following knee replacement [1].

Exposure

In the Femoral Triangle

1. A vertical incision is made just lateral to the pulse of the femoral artery and below the inguinal ligament (Fig. 16.1).
2. The incision is deepened through the superficial and deep fasciae.
3. Usually cutaneous branches of the femoral nerve can be encountered and followed deeply and proximally to the main stem of the nerve. If not, the femoral artery can be exposed first; the nerve is lateral to the artery outside the femoral sheath (Fig. 16.2).
4. The fascia overlying the psoas and iliacus muscles needs to be opened to expose the nerve.
5. The nerve is followed proximally under the inguinal ligament which can be opened.

In the Retroperitoneum
(Fig. 16.3; See Also Fig. 16.1, 15.2, and 15.3)

1. Open (described here) or laparoscopic.
2. Incision is made obliquely above the lateral half of the inguinal ligament.
3. The abdominal wall muscles can be split along their fibers: external oblique, internal oblique, and transversus abdominis.
4. The extraperitoneal fat is then encountered. Blunt dissection allows retracting the peritoneum with its contents medially exposing the retroperitoneal space.
5. The fascia overlying the iliacus and psoas muscles is opened.
6. The lateral femoral cutaneous nerve is first identified as it crosses in front of the iliacus muscle from the lateral border of the psoas towards the anterior superior iliac spine (ASIS).
7. The femoral nerve is found by opening the groove between the psoas major and iliacus muscles.

Electronic supplementary material The online version of this chapter 10.1007/978-3-319-14520-4_16 contains supplementary material, which is available to authorized users.

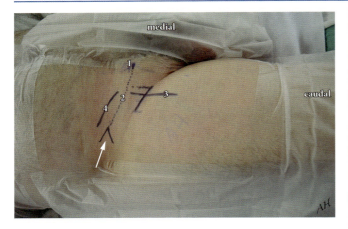

Fig. 16.1 Exposure of the right femoral nerve. The patient is positioned supine. *Arrow*: anterior superior iliac spine (ASIS); *1* pubic tubercle, *2* inguinal ligament (*dotted line*), *3* incision for exposure of the femoral nerve in the femoral triangle (thigh). It starts at the midpoint of the inguinal ligament, lateral to the femoral pulse. *4* incision for retroperitoneal exposure of the femoral nerve. A hip bump can be used to facilitate retroperitoneal exposure

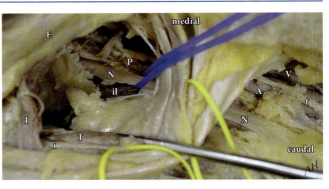

Fig. 16.3 Another cadaver specimen showing combined exposure of the right femoral nerve in the retroperitoneum and thigh. *E* external oblique, *I* internal oblique, *i* ilioinguinal nerve, *T* transversus abdominis, *P* psoas major, *il* iliacus, *N* femoral nerve, *A* femoral artery, *V* femoral vein, *f* femoral branch of genitofemoral nerve

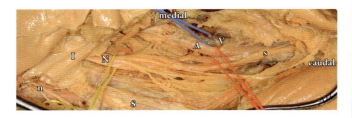

Fig. 16.2 Cadaveric dissection of the right femoral nerve in the thigh. *V* femoral vein, *A* femoral artery, *N* femoral nerve, *I* inguinal ligament, *n* lateral femoral cutaneous nerve, *S* sartorius, *s* saphenous nerve

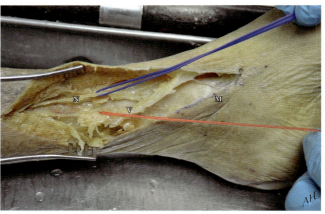

Fig. 16.4 Saphenous nerve exposure along the medial aspect of the left lower leg. *M* medial malleolus, *V* long saphenous vein, *N* saphenous nerve

The Saphenous Nerve

It is a small cutaneous branch that can be exposed in the superficial fascia, along the long saphenous vein anterior to the medial malleolus (Fig. 16.4), medial to the leg and knee joint. Followed proximally, it becomes deep to the sartorius and then accompanies the femoral artery in the adductor canal (Fig. 16.5).

Reference

1. Nahabedian MY, Johnson CA (2001) Operative management of neuromatous knee pain: patient selection and outcome. Ann Plast Surg 46(1):15–22

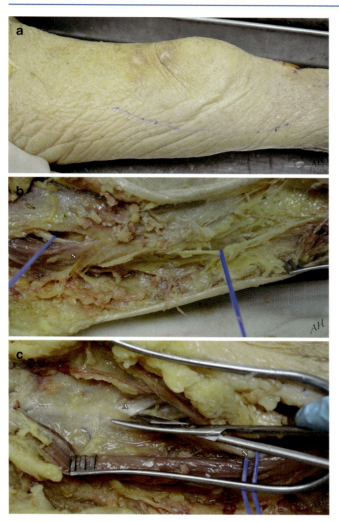

Fig. 16.5 (a) Saphenous nerve exposure along the medial aspect of the left knee. (b) The saphenous nerve (*blue loops*) is observed crossing deep to the sartorius (*S*). (c) The aponeurotic roof (*A*) of the adductor canal can be opened for further exposure. *S* sartorius, *N* saphenous nerve

Lateral Femoral Cutaneous Nerve 17

Surgical Anatomy

The lateral femoral cutaneous nerve (LFCN) arises from L2–L3 posterior divisions. It exits the psoas muscle along its lateral border and then runs obliquely caudally and laterally towards the anterior superior iliac spine (ASIS). It passes under the lateral part of the inguinal ligament to enter the thigh in its own sheath (a complete fascial tunnel surrounds the nerve). It supplies the skin of the anterolateral thigh.

Anatomical Variations

The nerve could be medial, superficial, or lateral to the ASIS.

Clinical Significance

Meralgia paresthetica is characterized by pain and paresthesias in the LFCN distribution. It can occur spontaneously or postoperatively after a prolonged prone position. It is more common in obese people (Fig. 17.1) [1].

Exposure

In the Thigh

1. Position: supine (Fig. 17.2).
2. Incision: vertical or oblique, about 1.5 cm medial to ASIS. The nerve can have a variable course and relationship with ASIS.
3. The incision is deepened through the superficial fascia.
4. The deep fascia overlying the sartorius muscle is opened exposing the LFCN (Fig. 17.3).
5. The nerve is followed proximally under the inguinal ligament, which is opened for full decompression (Fig. 17.4).
6. Due to the variability in the course of the nerve, ultrasound-assisted surgery is recommended, preferably with preoperative wire localization.
7. The author has recently adopted a nerve transposition, where the fascia underlying the nerve is opened allowing mobilization of the nerve medially for at least 2 cm. This eliminates the tight relationship between the nerve and the ASIS.

In the Retroperitoneum (Figs. 17.5 and 17.6)

See Chap. 16, femoral nerve in the retroperitoneum.

Electronic supplementary material The online version of this chapter 10.1007/978-3-319-14520-4_17 contains supplementary material, which is available to authorized users.

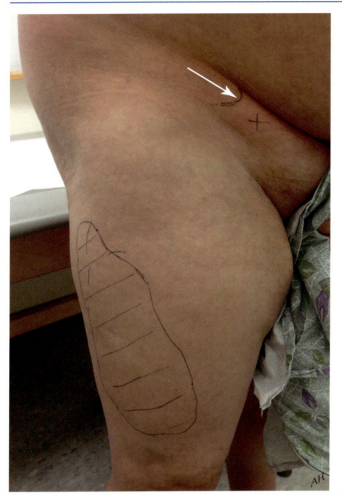

Fig. 17.1 A patient with right-sided meralgia paresthetica. The *parallel lines* mark the area of sensory changes in the anterolateral thigh. The "X" marks the location of a Tinel sign medial to the anterior superior iliac spine (ASIS) (*arrow*). Also note the size of the patient, which is typical for meralgia paresthetica

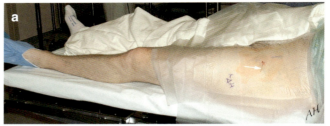

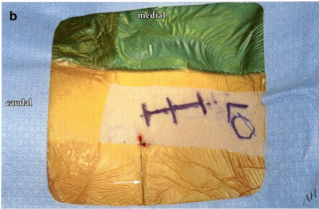

Fig. 17.2 (**a**) Another patient positioned for left lateral femoral cutaneous nerve decompression. The incision is made vertically 1–2 cm medial to the anterior superior iliac spine (ASIS). Note the guidewire (*white arrow*) placed to localize the nerve with preoperative ultrasound. (**b**) Close-up. The incision is marked medial to the ASIS (*blue arrow*), along the nerve trajectory mapped by preoperative ultrasound

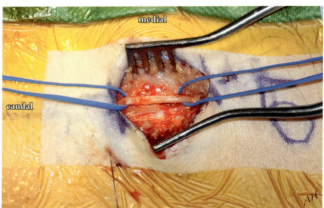

Fig. 17.3 The lateral femoral cutaneous nerve is exposed by opening the nerve sheath

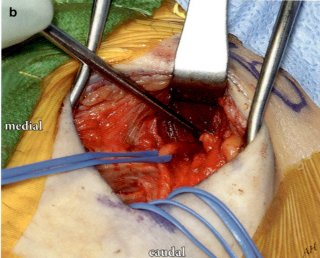

Fig. 17.4 (**a**) Proximally, the nerve is observed under the inguinal ligament (*arrows*), tethering it against the ASIS. (**b**) After full decompression, the nerve is completely released up to the retroperitoneal space

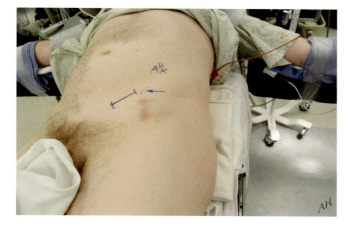

Fig. 17.5 Positioning and incision for retroperitoneal exposure of the left lateral femoral cutaneous nerve. The patient is supine with a hip bump. The incision is cranial and parallel to the inguinal ligament, medial to the ASIS (*blue arrow*). Note vertical incision from prior decompression caudal to ASIS

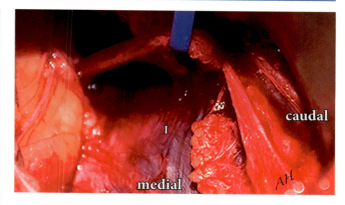

Fig. 17.6 The left lateral femoral cutaneous nerve (*blue loop*) is exposed in the retroperitoneum and transected. *I* iliacus

Reference

1. Cheatham SW, Kolber MJ, Salamh PA (2013) Meralgia paresthetica: a review of the literature. Int J Sports Phys Ther 8(6):883–893

Sciatic Nerve

Surgical Anatomy

The sciatic nerve has two major components: the tibial nerve from the anterior division of L4–S3 and the common peroneal nerve from the posterior division of L4–S2. It forms anterior to the piriformis muscle in the pelvis and exits the greater sciatic foramen usually below the piriformis, occasionally within the muscle, and then enters the thigh between the ischial tuberosity and the greater trochanter. Structures deep to the sciatic nerve from cranial to caudal are the gemellus superior, obturator internus, gemellus inferior, quadratus femoris, and adductor magnus. Structures superficial to the sciatic nerve include, in the buttock: gluteus maximus, and in the thigh: the long head of biceps femoris crosses the nerve from medial to lateral. The nerve supplies the hamstrings—biceps femoris, semimembranosus, and semitendinosus—as well as the ischial part of adductor magnus. The nerve ends by splitting into tibial and common peroneal (common fibular) components in the lower thigh.

Clinical Significance

The sciatic nerve is most vulnerable around the hip, where it can be injured with hip dislocations and hip or pelvic fractures or stretched during hip arthroplasty [1]. The peroneal component is more frequently and more severely involved. It is also the less likely to recover [2]. The last branch of the peroneal component in the thigh is the short head of the biceps femoris. This is important for lesion localization.

Electronic supplementary material The online version of this chapter 10.1007/978-3-319-14520-4_18 contains supplementary material, which is available to authorized users.

Exposures

In the Buttock

Transgluteal

1. Position: prone or lateral on a beanbag.
2. Incision: oblique, along the fibers of the gluteus maximus muscle (Fig. 18.1).
3. The incision is deepened through the superficial and deep fasciae.
4. The gluteus maximus muscle is split along its fibers. This can be done using Bovie coagulation or scissors with the vessels crossing coagulated using the bipolar forceps. A self-retaining retractor (Gelpi or cerebellar) is placed.
5. The fat layer deep to the gluteus maximus is then identified (Fig. 18.2). Palpation through the fat pad will reveal the big cord-like sciatic nerve.
6. Vertical dissection along the course of the sciatic nerve allows identification of the nerve (Fig. 18.3) with the posterior cutaneous nerve of the thigh superficial to it.
7. The nerve can be followed proximally to the piriformis muscle.
8. To treat a piriformis syndrome, a segment of the muscle should be transected as well as its nerve supply, which is deep to the muscle.
9. The pudendal nerve can be accessed through the same approach. It is deeper and more medial to the sciatic nerve.
10. For a fascicular biopsy, the nerve sheath is opened where abnormality is suspected based on preoperative MRI. An interfascicular dissection is performed. A grossly abnormal fascicle should be a target for the biopsy; otherwise a randomly selected fascicle is taken, preferably after silent stimulation by EMG. The fascicle is dissected cranially and caudally until a good size (4–6 cm) is obtained.

Gluteal Flap

1. Position: prone or lateral on a beanbag.
2. Question mark incision, starting along the lateral aspect of the hip, around the buttock, and then vertically along the mid-upper thigh (Figs. 18.1 and 18.4).
3. The incision is deepened through the superficial and deep fasciae.
4. The gluteus maximus is cut along its insertion leaving a cuff to stitch to (Fig. 18.5).
5. The gluteus maximus is then lifted up as a flap and reflected medially.
6. This exposes the next muscle layer: gluteus medius, gluteus minimus, and piriformis.
7. The fat pad below the piriformis is identified. Palpation in the fat pad will reveal the big cord-like sciatic nerve.
8. Vertical dissection along the course of the sciatic nerve allows identification of the nerve with the posterior cutaneous nerve of the thigh superficial to it and the pudendal nerve deep and medial to it.
9. The bed of the sciatic nerve (deep structures), from cranial to caudal, is made of gemellus superior, obturator internus, gemellus inferior, quadratus femoris, and adductor magnus (Fig. 18.6).
10. The nerve can be followed proximally towards the piriformis muscle.
11. The nerve can also be followed distally until it is crossed by the long head of the biceps femoris (Fig. 18.7).

In the Thigh

1. Position: prone.
2. Posterior midline thigh incision (Fig. 18.1).
3. Incision is deepened through superficial and deep fasciae.
4. The hamstrings are separated with the biceps femoris laterally and the semitendinosus and semimembranosus medially (Fig. 18.8). A self-retaining retractor is placed.
5. The fat layer deep to the muscles is then identified. Palpation through the fat pad will reveal the big cord-like sciatic nerve.
6. Vertical dissection along the course of the sciatic nerve allows identification of the nerve.
7. The nerve can be followed distally until its bifurcation.
8. It can be followed proximally until crossed by the long head of biceps femoris and then under the gluteus maximus.

Fig. 18.1 Different incisions for left sciatic nerve exposure. *1* transgluteal; *2* gluteal flap; *3* posterior thigh. This patient is positioned lateral. Alternatively, this can be done in a prone position

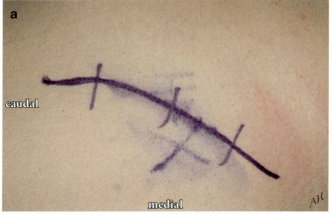 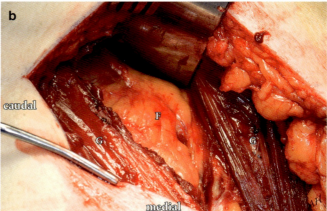

Fig. 18.2 (**a**) The incision is marked for a left transgluteal approach. (**b**) Separation of the gluteus maximus fibers (*G*) allows identification of the fat pad (*F*) underneath

Exposures

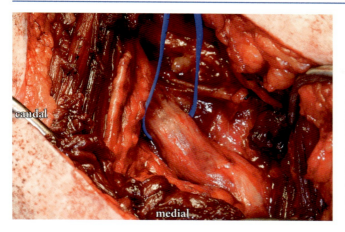

Fig. 18.3 The sciatic nerve (*blue loop*) is found within the fat pad

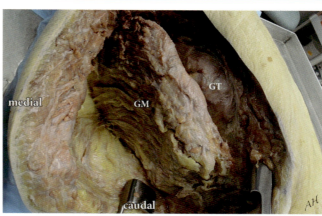

Fig. 18.5 The gluteus maximus (*GM*) flap is elevated. *GT* greater trochanter

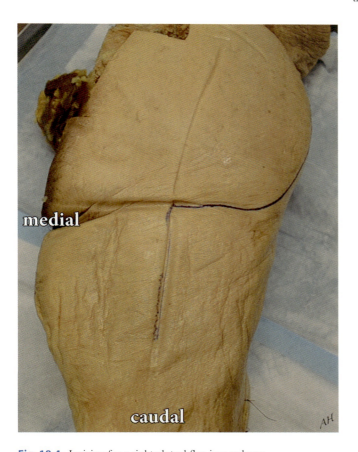

Fig. 18.4 Incision for a right gluteal flap in a cadaver

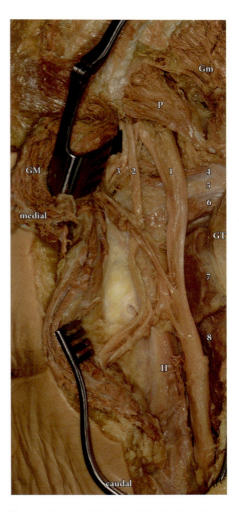

Fig. 18.6 The sciatic nerve (*1*) is found in the fat pad deep to the gluteus maximus (*GM*). *Gm* gluteus minimus, *P* piriformis, *GT* greater trochanter, *H* hamstrings; *2* posterior cutaneous nerve of the thigh; *3* pudendal nerve; *4–8* sciatic nerve bed: *4* gemellus superior; *5* obturator internus; *6* gemellus inferior; *7* quadratus femoris; *8* adductor magnus

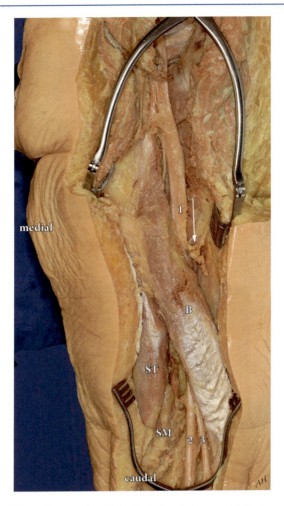

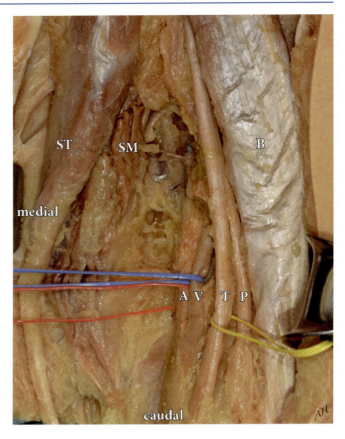

Fig. 18.7 In the posterior thigh, the right sciatic nerve (*1*) is crossed by the long head of biceps femoris (*B*) and then divides in the popliteal fossa into tibial (*2*) and common peroneal (common fibular) (*3*) nerves. *ST* semitendinosus, *SM* semimembranosus; *arrow* peroneal branch to the short head of biceps, last branch in the thigh

Fig. 18.8 Sciatic nerve in the right popliteal fossa. *ST* semitendinosus, *SM* semimembranosus, *B* biceps, *P* common peroneal (common fibular) nerve, *T* tibial nerve, *V* popliteal vein, *A* popliteal artery

References

1. Lee PT, Lakstein DL, Lozano B, Safir O, Backstein J, Gross AE (2014) Mid-to long-term results of revision total hip replacement in patients aged 50 years or younger. Bone Joint J 96-B(8): 1047–1051
2. Burks SS, Levi DJ, Hayes S, Levi AD (2014) Challenges in sciatic nerve repair: anatomical considerations. J Neurosurg 121(1): 210–218

Tibial Nerve

Surgical Anatomy

The tibial nerve separates from the sciatic nerve at the lower third of the thigh, the cranial part of the popliteal fossa. It continues in the midline, posterior to the popliteal vessels. In the upper leg, it passes between the two heads of the gastrocnemius and then under the soleal arch. It supplies all the muscles of the calf and a cutaneous contribution to the sural (medial sural cutaneous) nerve. It then appears in the medial aspect of the ankle, where it gives off a calcaneal branch, enters the tarsal tunnel, and divides into medial and lateral plantar nerves. These enter their individual tunnels under the abductor hallucis and supply the intrinsic muscles of the foot and the skin of the sole of the foot.

Clinical Significance

The tibial nerve can be entrapped at the soleal arch or the tarsal tunnel. Morton's neuroma affects the digital branches of the medial plantar nerve, either between the second and third or the third and fourth toes.

The tibial nerve can be injured with knee dislocation and/or fractures. It usually recovers much better and faster than the peroneal nerve.

Exposures

In the Popliteal Fossa and Upper Leg (Fig. 19.1)

1. Position: prone.
2. Incision in the posterior midline of the lower thigh and upper leg. If the knee crease is to be crossed, this should be done obliquely or by a Z-shaped incision.
3. The superficial and deep fasciae are opened.
4. The tibial nerve is found in the midline, while the common peroneal nerve can be found in the lateral aspect along the biceps femoris.
5. The tibial nerve can be followed distally, where it gives off multiple branches to the calf muscles as well as a contribution to the sural nerve. It passes between the two heads of gastrocnemius and then under the soleal arch. The latter could be split open for decompression of the tibial nerve.

In the Tarsal Tunnel (Fig. 19.2) [1]

1. Position: supine with the hip externally rotated and the knee slightly flexed
2. A tourniquet can be used.
3. A curvilinear incision is made midway between the medial malleolus and the calcaneus.
4. The incision is deepened through the superficial and deep fasciae.
5. The flexor retinaculum is opened sharply. This allows exposing the tibial nerve and posterior tibial vessels.
6. The calcaneal branch arises from the posterior aspect of the tibial nerve and runs posteriorly and caudally.
7. In the distal part of the tarsal tunnel, the tibial nerve splits into its two terminal branches: medial and lateral plantar nerves.
8. These are followed distally under the abductor hallucis.
9. They enter individual tunnels. These should be individually decompressed, and then the septum between the two tunnels should be resected.

Electronic supplementary material The online version of this chapter 10.1007/978-3-319-14520-4_19 contains supplementary material, which is available to authorized users.

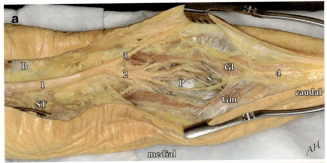

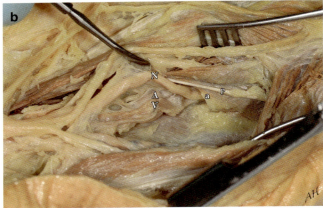

Fig. 19.1 (**a**) Sciatic (*1*), tibial (*2*), and common peroneal (common fibular) (*3*) nerves exposure in the right popliteal fossa. Position: prone. *B* biceps, *ST* semitendinosus, *Gm* gastrocnemius, medial head, *Gl* gastrocnemius, lateral head, *S* soleus, *P* popliteus *4* sural nerve with contributions from both the tibial and common peroneal nerves. (**b**) Close-up view showing the popliteal vein (*V*), popliteal artery (*A*), and tibial nerve (*N*) exiting the popliteal fossa under the soleal arch (*a*). *p* plantaris tendon

In the Foot (for Morton's Neuroma) (Fig. 19.3)

1. Position: supine.
2. Incision on the dorsum of the foot between the second and third toes or the third and fourth.
3. Dissection is carried deeply until the intermetatarsal ligament is transected.
4. The digital nerve is thus decompressed.
5. The neuroma could be left alone, resected, or resected followed by nerve grafting.

Complications of Tarsal Tunnel Release

1. Pain from injury to the saphenous nerve anteriorly or the calcaneal branch posteriorly.
2. Residual symptoms from inadequate distal decompression.
3. Vascular injury to the posterior tibial vessels.
4. Wound infection or dehiscence especially in diabetic patients.

Complications of Tarsal Tunnel Release

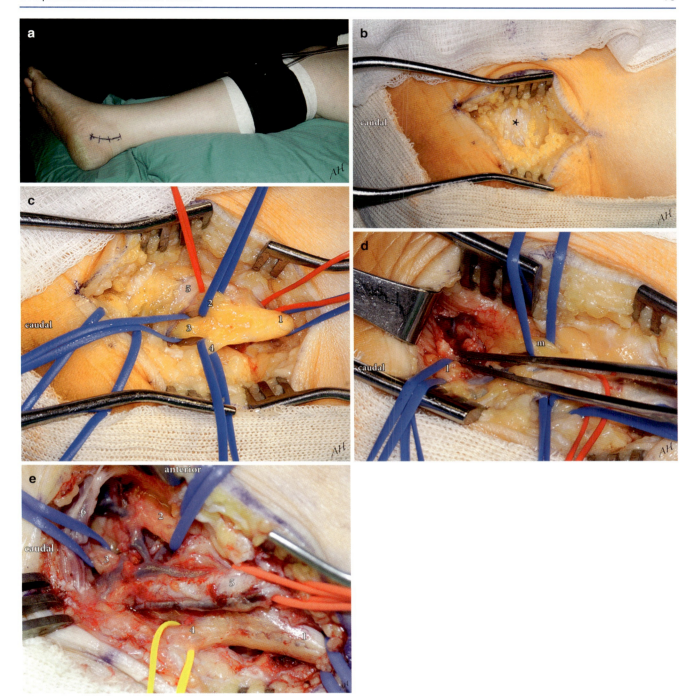

Fig. 19.2 Right tarsal tunnel release. (**a**) The patient is positioned supine, with the hip externally rotated and the knee slightly flexed. A tourniquet can be used. Incision is made between the medial malleolus and the calcaneus. (**b**) The flexor retinaculum (*) is exposed then opened. (**c**) The tibial (*1*), medial plantar (*2*), and lateral plantar (*3*) nerves are exposed, as well as the calcaneal branch (*4*). The posterior tibial vessels (*5*) are also dissected free. (**d**) The medial (*m*) and lateral (*l*) plantar nerves are followed distally into their individual tunnels under the abductor hallucis muscle. The septum in between (held with a forceps) needs to be resected for a full decompression. (**e**) Final view after tarsal tunnel release. *1* tibial nerve, *2* medial plantar nerve, *3* lateral plantar nerve, *4* calcaneal branch, *5* posterior tibial vessels, *6* abductor hallucis

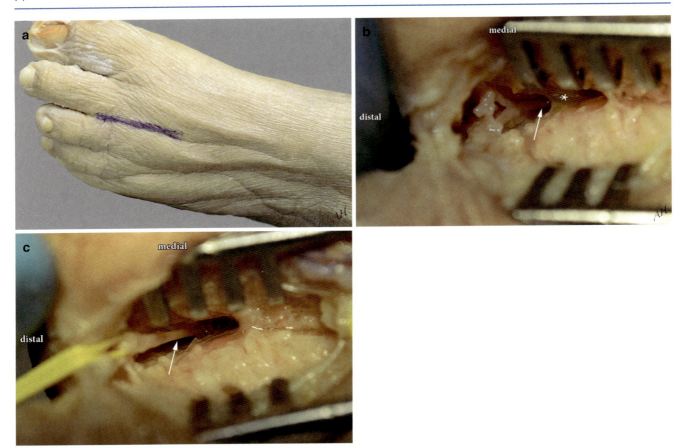

Fig. 19.3 Cadaveric illustration of surgery for left Morton's neuroma. (**a**) Incision is made between the second and third toes or the third and fourth depending on the location of the neuroma. (**b**) The plantar digital nerve (*arrow*) is observed deep to the intermetatarsal ligament (*). (**c**) The nerve (*arrow*) is free after decompression

Reference

1. Dellon AL (2008) The Dellon approach to neurolysis in the neuropathy patient with chronic nerve compression. Handchir Mikrochir Plast Chir 40(6):351–360

Common Peroneal Nerve (aka Common Fibular Nerve)

Surgical Anatomy

The common peroneal nerve (also called common fibular nerve) is one of the two terminal branches of the sciatic nerve in the lower third of the thigh. It runs laterally, medial to the biceps femoris tendon, then posterior to the fibular head, and lateral to the fibular neck. It gives off the sural communicating branch (lateral sural cutaneous nerve). This typically joins the contribution from the tibial nerve to form the sural nerve. The latter descends along the lateral aspect of the leg anterior to the Achilles tendon. The lateral dorsal cutaneous nerve is one of its terminal branches. The common peroneal nerve then divides into superficial peroneal, deep peroneal, and articular (recurrent genicular) branches.

The superficial peroneal nerve supplies the peroneus longus and peroneus brevis muscles. It gives off two terminal cutaneous branches: medial dorsal cutaneous and intermediate dorsal cutaneous nerves. These pierce the deep fascia at the lower third of the leg and supply the dorsal surface of the foot and toes except the first web space and the lateral aspect of the little toe.

The deep peroneal nerve supplies the tibialis anterior, extensor hallucis longus, extensor digitorum longus, and peroneus tertius. It continues under the extensor retinaculum in the anterior tarsal tunnel and then divides in the dorsum of the foot into a lateral muscular branch supplying the extensor digitorum brevis and a medial cutaneous branch supplying the adjacent sides of the first and second toes.

Electronic supplementary material The online version of this chapter 10.1007/978-3-319-14520-4_20 contains supplementary material, which is available to authorized users

Clinical Significance

The common peroneal nerve is easily injured with knee dislocations and/or fractures. It is slow to recover, and the prognosis is generally poor especially with longer lesions. Injury can result in a foot drop with intact inversion (Fig. 20.1). The recurrent genicular branch supplies the superior tibiofibular joint. Inflammation of this joint can result in retrograde synovial filling of the articular branch and the peroneal nerve, with development of an intraneural ganglion cyst (Chap. 26).

The cutaneous branches of the superficial peroneal nerve can be injured during fasciotomies for compartment syndrome. Different cutaneous branches are also vulnerable during ankle surgery [1].

Exposures

In the Popliteal Fossa

1. Position: prone.
2. Incision: oblique along the posterior border of the biceps femoris.
3. The superficial and deep fasciae are opened.
4. The common peroneal nerve is found within the popliteal fossa fat medial to the biceps femoris tendon (see Fig. 19.1).

Around the Fibular Neck (Fig. 20.2)

1. Position: lateral on a beanbag.
2. Incision: curvilinear posterior to the fibular head and lateral to the fibular neck.
3. The superficial and deep fasciae are opened. The common peroneal nerve can be felt against the head of fibula.
4. The nerve sheath is opened sharply using Metzenbaum scissors.

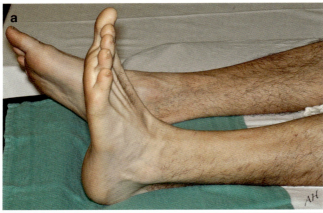
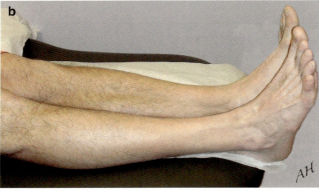

Fig. 20.1 Pre- (a) and postoperative (b) photos from a patient with right foot drop from common peroneal nerve entrapment at the fibular neck. *Arrow*: scar from peroneal decompression surgery

5. The nerve is followed proximally towards the popliteal fossa and decompressed.
6. The nerve is followed distally under the peroneal fascia, which is opened for full decompression: first the fascia superficial to the muscle is opened, and then the muscle is retracted exposing the deeper fascial layer. This is opened as well.
7. This exposes the three branches of the peroneal nerve: superficial peroneal, deep peroneal, and articular branches.

In the Lower Leg and Anterior Ankle

Superficial Peroneal (Fig. 20.3)
1. Position: supine.
2. Oblique incision along the lateral aspect of the leg and dorsum of the ankle.
3. The terminal dorsal cutaneous branches can be found piercing the deep fascia in the lower third of the leg. Then they course in the subcutaneous fat with the medial branch medially and the intermediate branch laterally.
4. Ultrasound assistance could be used to localize these small cutaneous nerves.

Deep Peroneal (Anterior Tarsal Tunnel) (Fig. 20.4)
1. Position: supine.
2. Incision: midline dorsum of the ankle.
3. The extensor retinaculum is opened.
4. The deep peroneal nerve is found with the anterior tibial vessels. Tibialis anterior and extensor hallucis longus tendons are medial to it. Extensor digitorum longus is lateral to it.
5. The extensor retinaculum separates the superficial peroneal nerve branches in the subcutaneous fat from the deep peroneal nerve in the anterior tarsal tunnel.

Sural Nerve (Fig. 20.5)

1. Position: depends on the indication. For a nerve biopsy, the patient is positioned lateral under local anesthesia and sedation. For nerve grafting, the position is dictated by the primary nerve to be repaired. Sural nerve can even be harvested in a supine position with the leg elevated by an assistant.
2. Incision is made lateral and anterior to the Achilles tendon.
3. Once a self-retaining retractor is placed, the short saphenous vein can be seen within the subcutaneous fat. The sural nerve runs with the short saphenous vein.
4. Dissection is carried vertically along the nerve, to free it all around.
5. The nerve can be followed proximally until its origin in the upper leg from both the tibial and the peroneal nerves. Occasionally these two contributions to the sural nerve remain separate and continue as two distinct nerves.

Complications of Sural Nerve Harvest

1. Loss of sensation on the lateral aspect of the dorsum of the foot and little toe
2. Pain, from neuroma formation
3. Wound infection

Complications of Sural Nerve Harvest

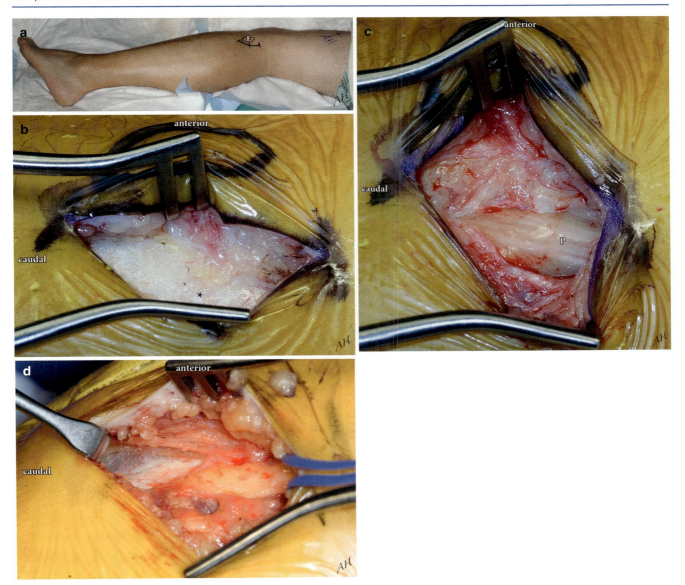

Fig. 20.2 Another patient presenting for left common peroneal nerve decompression at the fibular neck. (**a**) Position, lateral on a beanbag; incision, posterior to the fibular head (F) and lateral to the fibular neck. (**b**) The nerve sheath is exposed (*) and opened (**c**), thus exposing the common peroneal (common fibular) nerve (*P*). The fascia overlying the peroneal muscles is exposed (**d**) and opened (**e**). (**f**) The muscle is retracted, thus exposing the fascia deep to it with its sharp edge (*arrowheads*) compressing the nerve. (**g**) This fascia is opened achieving full distal decompression. The three terminal branches of the common peroneal nerve are exposed: *S* superficial peroneal; *D* deep peroneal; and *A* articular. (**h**) Full proximal decompression medial to the tendon of the biceps femoris (*B*)

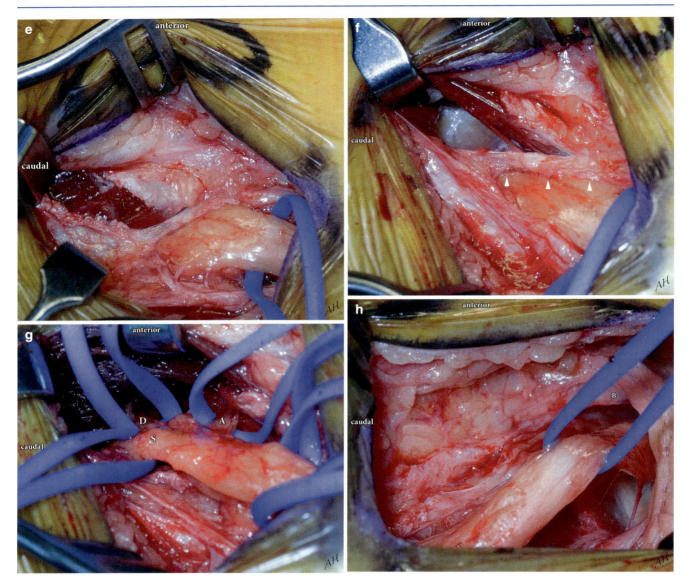

Fig. 20.2 (continued)

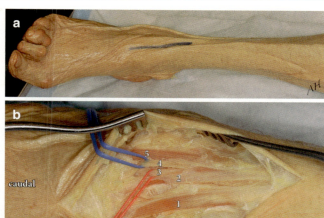

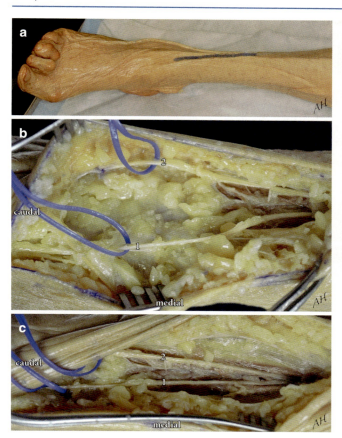

Fig. 20.3 (a) Incision for exposure of the cutaneous branches of the right superficial peroneal (superficial fibular) nerve. The medial (*1*) and intermediate (*2*) dorsal cutaneous nerves are observed in the subcutaneous fat distally (b) and deep to the peroneal fascia proximally (c)

Fig. 20.4 (a) Incision for anterior tarsal tunnel release. (b) Once the extensor retinaculum is opened, the following structures can be identified from medial to lateral: *1* tibialis anterior, *2* extensor hallucis longus, *3* anterior tibial artery, *4* deep peroneal (deep fibular) nerve, *5* extensor digitorum longus

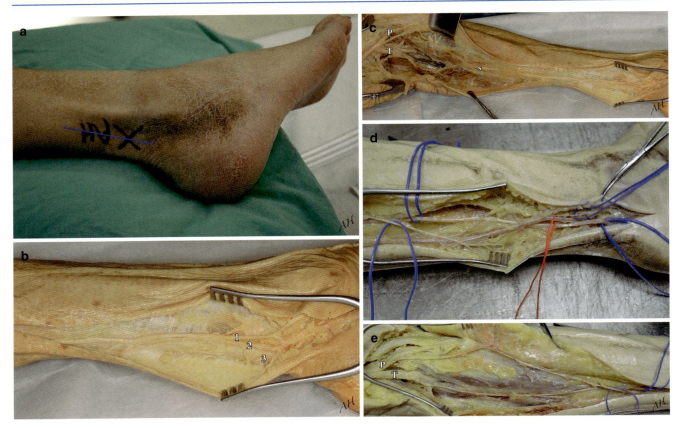

Fig. 20.5 Sural nerve exposure. (**a**) Position is lateral, and the incision is made laterally, anterior to the Achilles tendon. (**b**) The skin is opened; within the fat is the sural nerve (*1*), along the short saphenous vein (*2*), anterior to the Achilles tendon (*3*). (**c**) As needed, the sural nerve (*s*) can be followed proximally until its origins from the tibial (*T*) and common peroneal (*P*) nerves. (**d, e**) Occasionally (different specimen), the two components of the sural nerve never meet and remain as two distinct nerves (*blue loops*) until their origins from the tibial (*T*) and common peroneal (*P*) nerves. *Red loop* short saphenous vein

Reference

1. Duscher D, Wenny R, Entenfellner J, Weninger P, Hirtler L (2014) Cutaneous innervation of the ankle: an anatomical study showing danger zones for ankle surgery. Clin Anat 27:653–658

Obturator Nerve

21

Surgical Anatomy

The obturator nerve arises from the anterior division of L2–L4 ventral rami. It courses medial to the psoas major muscle, along the sidewall of the pelvis, with the internal iliac vessels. It exits the pelvis through the obturator canal in the upper part of the obturator foramen, where it divides into anterior and posterior divisions. The anterior division passes above the obturator externus and then in between the adductor brevis posteriorly and pectineus and the adductor longus anteriorly. The posterior division passes through the obturator externus and then between the adductor brevis and the adductor magnus. The anterior division supplies the adductor longus, adductor brevis, and gracilis muscles. The posterior division supplies the obturator externus and adductor part of the adductor magnus. They also have articular branches and cutaneous branches that supply the medial aspect of the thigh.

Clinical Significance

The gracilis branch is very important when harvesting the gracilis muscle for a free flap with its neurovascular pedicle [1].

Anterior obturator neurectomy and adductor myotomy are used to treat spasticity of the adductor muscles especially in children with cerebral palsy [2]. Intrapelvic obturator neurectomy has also been used.

The obturator nerve can be used as donor for nerve transfers for femoral nerve palsy [3].

Electronic supplementary material The online version of this chapter 10.1007/978-3-319-14520-4_21 contains supplementary material, which is available to authorized users.

Exposure

Intrapelvic (Fig. 21.1; See Also Fig. 15.3)

1. Open or laparoscopic.
2. General surgeon for access.
3. Horizontal Pfannenstiel incision can be used.
4. Once the abdomen is entered, the nerve is identified along the sidewall of the pelvis. Electrical stimulation is used for confirmation (EMG).
5. If neurectomy is the goal, a segment of the nerve is then resected.

In the Thigh

1. Position: supine with the hip abducted and externally rotated.
2. Incision is made along the adductor longus tendon up to the pubic tubercle (Fig. 21.2).
3. The superficial and deep fasciae are opened.
4. The adductor longus and pectineus are identified.
5. The adductor longus can be resected first if myectomy is part of the procedure. The anterior division of the obturator nerve can then be found lying anterior to the adductor brevis muscle.
6. If no myectomy is performed, the plane between the adductor longus and pectineus is entered (Fig. 21.3a). These two muscles are separated. The anterior division of the obturator nerve is found deep to these two muscles (Fig. 21.3b) and superficial to the adductor brevis.
7. To find the posterior division, the adductor brevis can be retracted and the nerve is found between the adductor brevis and magnus (Fig. 21.4).
8. The plane between the adductor longus and gracilis can be entered to isolate the gracilis branch of the anterior division as well as its vascular pedicle (Fig. 21.5).

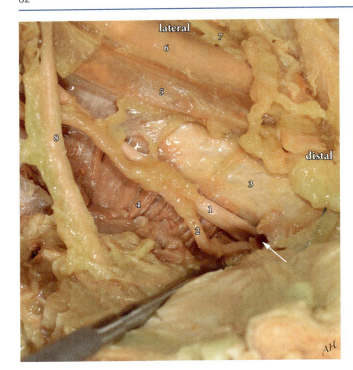

Fig. 21.2 Obturator nerve exposure in the right thigh. Incision is made along the adductor longus

Fig. 21.1 Intrapelvic exposure of the left obturator nerve. (*1*) obturator nerve; the nerve exits the pelvis through the obturator canal (*arrow*), after it divides into anterior and posterior divisions; (*2*) obturator artery; (*3*) superior pubic ramus; (*4*) obturator internus; (*5*) external iliac vein; (*6*) external iliac artery; (*7*) genital branch of the genitofemoral nerve; (*8*) ureter

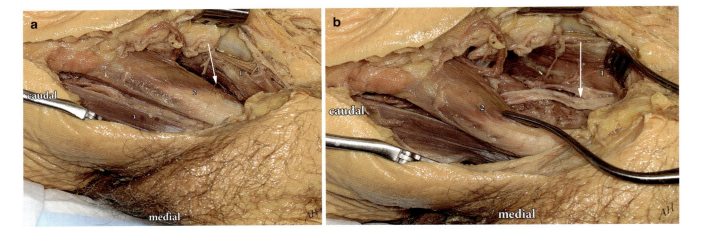

Fig. 21.3 (**a**) The plane (*arrow*) between the pectineus (*1*) and adductor longus (*2*) is entered to find the anterior division of the obturator nerve. *3* gracilis. (**b**) Once the pectineus (*1*) and adductor longus (*2*) are separated, the anterior division of the obturator nerve (*arrow*) can be found

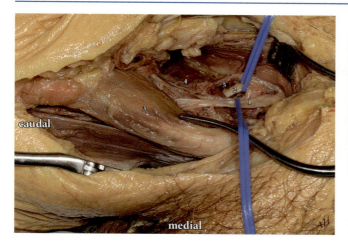

Fig. 21.4 The anterior division of the right obturator nerve (*1*) is observed between the adductor longus (*3*) and adductor brevis (*4*). The posterior division (*2*) is between the adductor brevis (*4*) and adductor magnus (*5*)

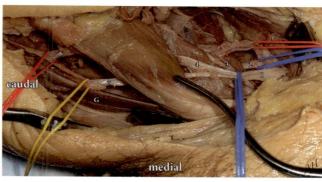

Fig. 21.5 The gracilis (*G*) neurovascular pedicle is dissected. (*1*) anterior division of the obturator nerve. (*2*) posterior division of the obturator nerve; (*3–5*) branches of the anterior division; (*3*) to adductor brevis; (*4*) to adductor longus; (*5*) to gracilis. (*6*) gracilis vascular pedicle

References

1. Doi K, Sakai K, Kuwata N, Ihara K, Kawai S (1995) Double free-muscle transfer to restore prehension following complete brachial plexus avulsion. J Hand Surg Am 20(3):408–414
2. Snela S, Rydzak B (2002) The value of the adductor tenotomy with obturator neurectomy in the treatment of the hips at cerebral palsy children. Early clinical and radiological examination results. Orthop Traumatol Rehabil 4(1):11–14
3. Campbell AA, Eckhauser FE, Belzberg A, Campbell JN (2010) Obturator nerve transfer as an option for femoral nerve repair: case report. Neurosurgery 66(ons2):E375

Ilioinguinal Nerve

Surgical Anatomy

The ilioinguinal nerve arises from the ventral ramus of L1. It emerges from the lateral aspect of the psoas muscle cranial to the lateral femoral cutaneous nerve. It courses in the retroperitoneum and then pierces the transversus abdominis at the lateral part of the iliac crest to course between the transversus abdominis and internal oblique muscles. More anteriorly, it pierces the internal oblique to lie between this muscle and the external oblique, below the spermatic cord (or round ligament of the uterus). It supplies the skin of the medial groin, medial upper thigh, upper scrotum, and root of the penis (mons pubis and labium majus in females) as well as the abdominal wall muscles.

Clinical Significance

The ilioinguinal nerve can be injured during hernia repair whether open or laparoscopic, especially with mesh placement. It can also be injured with lower abdominal incisions for other reasons like Caesarean section or female pelvic pathologies [1]. This may cause severe debilitating pain. If refractory to medical management, proximal neurectomy is a good surgical option [2].

Exposure

1. Position: supine, with or without a hip bump.
2. Ultrasound assistance could be used to find the nerve.
3. Incision should be parallel to and just above the anterior part of the iliac crest. It should extend anteriorly above the inguinal ligament (Fig. 22.1).
4. The superficial fatty layer (Camper's fascia) and deep fibrous layer (Scarpa's fascia) are opened.
5. The external oblique muscle is opened along its fibers. Medially, the nerve can be found between the external and internal oblique muscles (Fig. 22.2)
6. The internal oblique is opened along its fibers. Self-retaining retractors are placed.
7. Laterally, the nerve is found between the internal oblique and the transversus abdominis muscles (Fig. 22.3), coursing parallel to the iliac crest, usually with the deep circumflex iliac vessels.
8. Scar tissue can be released or a segment of the nerve can then be transected, as indicated.
9. Of note, the iliohypogastric nerve can be in close proximity to the ilioinguinal nerve. If transection is planned, they should be both transected.

Electronic supplementary material The online version of this chapter 10.1007/978-3-319-14520-4_22 contains supplementary material, which is available to authorized users.

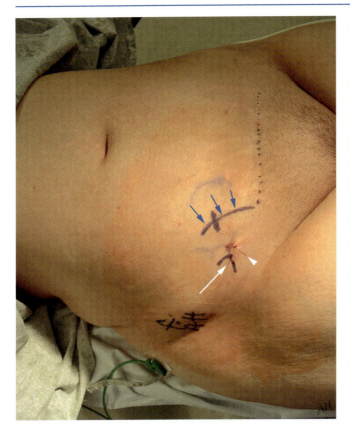

Fig. 22.1 The patient had a painful scar from previous Caesarean section (*dotted line*). She is positioned supine for exposure and transection of the right ilioinguinal nerve. The incision (*blue arrows*) is made parallel and above the inguinal ligament. *White arrow*: anterior superior iliac spine (ASIS). *Arrowhead*: guidewire placed with a preoperative ultrasound for localization

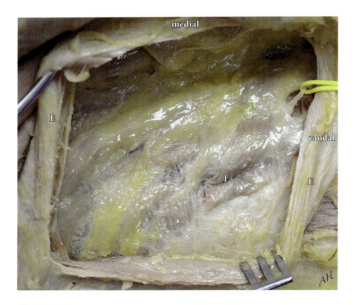

Fig. 22.2 In a cadaveric dissection, the external oblique aponeurosis (*E*) is opened, revealing the ilioinguinal nerve (*yellow loop*) between it and the internal oblique muscle (*I*) towards the medial part of the exposure

Fig. 22.3 More laterally, the ilioinguinal nerve (*yellow loop*) is found between the internal oblique (*I*) and transversus abdominis muscles (*T*). *E* external oblique

References

1. Stulz P, Pfeiffer KM (1982) Peripheral nerve injuries resulting from common surgical procedures in the lower portion of the abdomen. Arch Surg 117(3):324–327
2. Kim DH, Murovic JA, Tiel RL, Kline DG (2005) Surgical management of 33 ilioinguinal and iliohypogastric neuralgias at Louisiana State University Health Sciences Center. Neurosurgery 56(5):1013–1020

Part IV

Technical Notes

Trauma

Grading of Nerve Injuries

Seddon Classification 1941 (Terms by Henry Cohen) [1]

- Neurapraxia: Conduction block with minimal anatomical disruption
- Axonotmesis: Axonal disruption
- Neurotmesis: Complete nerve discontinuity

Sunderland Classification 1951 [2]

- Grade I: Neurapraxia
- Grade II: Axonotmesis, nerve sheaths intact
- Grade III: Axonotmesis, endoneurial disruption
- Grade IV: Axonotmesis, perineurial disruption
- Grade V: Neurotmesis

Timing of Nerve Repair

1. Sharp injuries usually result in neurotmesis and require immediate repair.
2. Blunt injuries, such as stretch or compression injuries, as well as gunshot wounds usually result in incomplete disruption of the nerve. They are either graded as axonotmesis or neurapraxia and should be treated conservatively for potentially spontaneous recovery. Surgery is indicated if they don't recover by 3–6 months. Avulsion injuries were traditionally treated conservatively for 3–6 months but now there is a trend to operate early since these are neurotmetic lesions.
3. Open crush injuries with damaged nerve endings are tagged initially. Repair is delayed for about 3 weeks. To tag a nerve, we use colored nonabsorbable sutures like Prolene or Nylon and hemoclips with slight tension to avoid retraction. This type of injury is rare.

Suturing

1. Magnification is recommended using microscope or loupes.
2. Nonabsorbable sutures: Prolene or Nylon. Size varies from 5/0 to 9/0 depending on the size of the nerve.
3. Avoid tension. To gain length, one may circumferentially dissect the nerve proximally and distally for better mobilization.
4. Minimize manipulation of the fascicles; as much as possible use fine pickups to hold the nerve sheaths.
5. Trim the nerve endings to see sprouting fascicles. Frozen section could be used to confirm suturing to viable nerve endings.
6. Use the minimum number of interrupted sutures to coapt the ends, typically 2, 180° from each other. Flat knots are preferred, usually 3 or 4 knots. The knots are tied tight enough to coapt the ends, but not too tight to avoid crushing the nerve endings. They do not need to be watertight.
7. May use a tube to wrap around, and glue on the surface of the tube and the interface with the nerve.

Electronic supplementary material The online version of this chapter 10.1007/978-3-319-14520-4_23 contains supplementary material, which is available to authorized users.

Direct Repair

1. Direct repair should always be attempted as long as it is not done under tension.
2. Attempt to match fascicles, by their sizes and/or the vessels in the nerve sheaths.
3. Suture technique described above.

Grafting (Figs. 23.1 and 23.2)

1. If direct repair is not possible, a nerve graft is used to bridge the gap.
2. Donors: autografts: sural nerve (Chap. 20), superficial radial nerve, or saphenous nerve. Allografts are now available.
3. Ensure the proximal stump is healthy by inspecting it, trimming as needed; and functional by using neuromonitoring with somatosensory evoked potentials and/or motor evoked potentials especially when dealing with supraclavicular brachial plexus injuries.

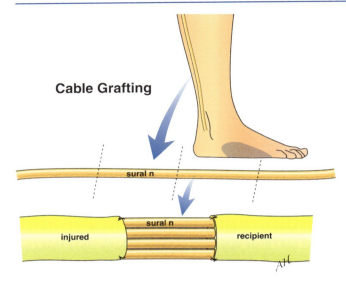

Fig. 23.1 Technique of cable grafting. A donor nerve (sural n) is harvested from the outer aspect of the leg anterior to the Achilles tendon. Several segments of nerve are cut (*dashed lines*) to match the length and diameter desired for the injured recipient nerve. This usually results in an area of sensory loss on the outer aspect of the dorsum of the foot (*shaded area*)

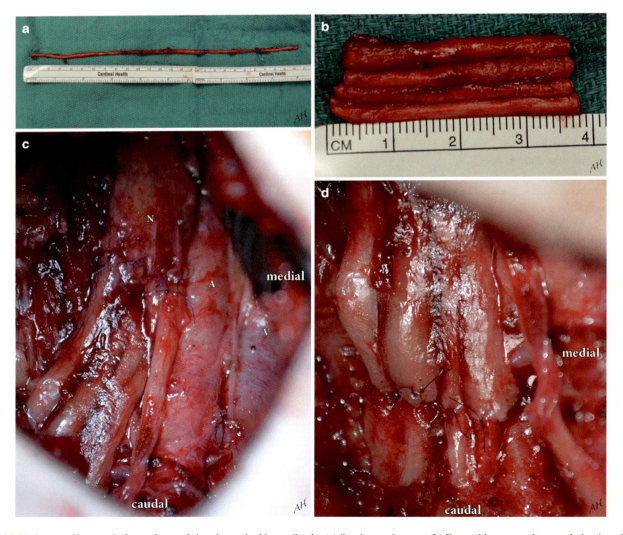

Fig. 23.2 A case of iatrogenic femoral nerve injury is repaired immediately. (**a**) Sural nerve harvest. (**b**) Four cables are used to match the size of the femoral nerve. (**c**) Proximal suture line with 5-0 Prolene. *V* femoral vein, *A* femoral artery, *N* femoral nerve, proximal stump. (**d**) Distal suture line

Tube Repair

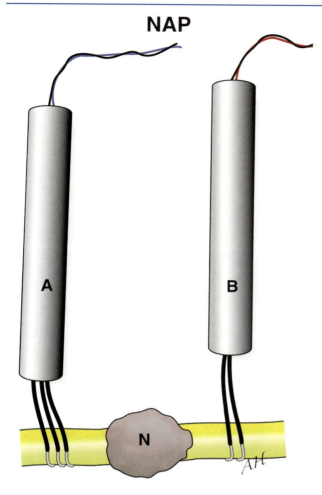

Fig. 23.3 A neuroma in continuity (*N*) along the course of an injured nerve can be tested for nerve action potentials (*NAPs*). Stimulating (*A*) and receiving (*B*) electrodes are placed on either side of the neuroma. The presence of NAPs signifies the presence of viable axons through the neuroma

4. Suture technique described above.
5. Nerve action potentials (NAPs) can be used to test a neuroma in continuity, intraoperatively, for the presence of functional axons (Fig. 23.3) [3]. Absent NAPs is an indication for graft repair.

Nerve Transfers (Fig. 23.4)

1. Nerve transfers borrow a branch or a fascicle from a normal nerve to reinnervate a paralyzed muscle.
2. Advantage: short distance to target and one suture line.

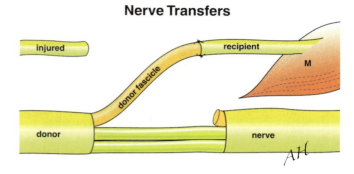

Fig. 23.4 Technique of nerve transfers. A fascicle is selected from a donor nerve and cut distally. The nonfunctional recipient nerve is cut proximally and sutured to the donor fascicle. This technique has the advantage of being close to the target muscle (*M*) and having one suture line, allowing shorter recovery time

3. Examples include ulnar fascicle to musculocutaneous (biceps brachii; Oberlin's transfer) [4], median fascicle to musculocutaneous (brachialis) (Figs. 23.5 and 23.6), radial branch to axillary (anterior division) (Fig. 23.7), spinal accessory to suprascapular, and intercostals to musculocutaneous.
4. For interfascicular dissection micropickups and tenotomy or microscissors are used to open the nerve sheaths.
5. Individual fascicles are separated gently.
6. Electrical stimulation and either inspection of the limb or needle electromyography are used to select redundant motor fascicles with no fine motor target (e.g., select the wrist and not the hand).
7. The donor fascicle is cut distally.
8. The target recipient is cut proximally.
9. Rarely if there is a gap between the donor and recipient nerves, an interposition graft is used, but this lowers the chances of regeneration.
10. Suture technique described above for end to end.
11. In case the recipient nerve has residual function, it is not transected, and the donor nerve is sutured end to side, by opening a slit in the epineurium of the recipient nerve.

Tube Repair

Occasionally for short gaps especially in digital nerves, a tube is used to bridge the gap. The nerve endings are sutured to the ends of the tube.

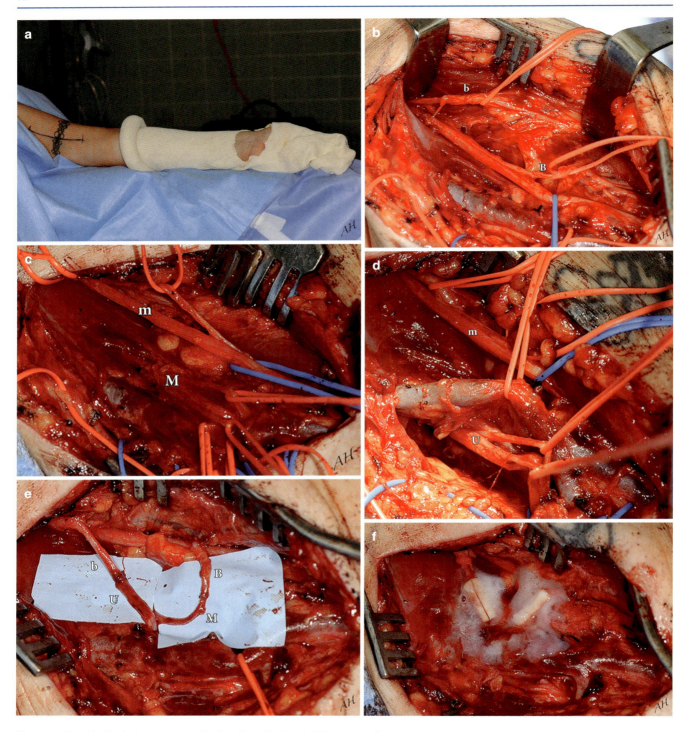

Fig. 23.5 Double fascicular nerve transfer for elbow flexion. (**a**) The patient is positioned supine with the arm abducted. (**b**) The musculocutaneous nerve (*m*) is exposed with its branches to the biceps brachii (*b*) and brachialis (*B*). These are the targets. (**c**) Interfascicular dissection of the median nerve (*M*) allows identification of fascicles that can be used as donors. EMG monitoring allows selection of fascicles with wrist function rather than hand function. (**d**) The process is repeated with the ulnar nerve (*U*). (**e**) Donors are cut distally, recipients proximally. An ulnar fascicle (*U*) is transferred to the biceps brachii branch (*b*) and a median fascicle (*M*) is transferred to the brachialis branch (*B*). (**f**) The suture sites can be supplemented with wrap tubes and glue

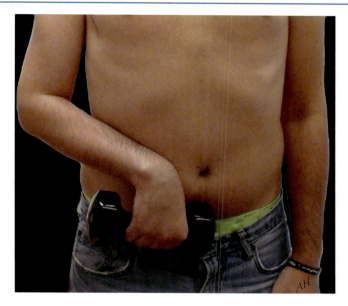

Fig. 23.6 One year after double transfer for elbow flexion, another patient is able to lift 8 lb

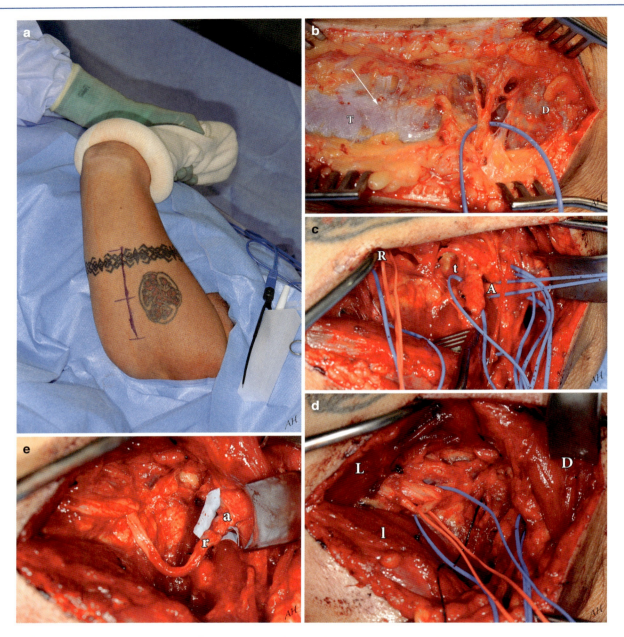

Fig. 23.7 Radial to axillary nerve transfer. (**a**) The patient is positioned supine with a shoulder bump or lateral on a beanbag. (**b**) At the posterior border of the deltoid (*D*), a cutaneous branch of the axillary nerve can be identified (*blue loop*) and followed proximally to find the axillary nerve. Note the white raphe (*arrow*) between the lateral and long heads of the triceps (*T*). This is the plane of the radial nerve. (**c**) The deltoid is retracted anteriorly and the axillary nerve (*A*) is found in the quadrangular space. The radial nerve (*R*) is observed more distally in the lower triangular space. The teres major tendon (*t*) separates the two nerves. (**d**) The radial nerve is found between the lateral (*L*) and long (*l*) heads of the triceps. A triceps branch (*red loop*) is selected and cut distally to use as a donor. (**e**) The anterior division of the axillary nerve (*a*) is cut proximally and sutured to a triceps branch of the radial nerve (*r*)

References

1. Seddon HJ (1942) A classification of nerve injuries. Br Med J 4260:237–239
2. Sunderland S (1951) A classification of peripheral nerve injuries producing loss of function. Brain 74:491–516
3. Robert EG, Happel LT, Kline DG (2009) Intraoperative nerve action potential recordings: technical considerations, problems, and pitfalls. Neurosurgery 65(4 Suppl):A97–A104
4. Oberlin C, Beal D, Leechavengvongs S, Salon A, Dauge MC, Sarcy JJ (1994) Nerve transfer to biceps muscle using a part of ulnar nerve for C5-C6 avulsion of the brachial plexus: Anatomical study and report of four cases. J Hand Surg 19A(2):232–237

Neuroma

Indications

The main indication for surgery is painful neuroma intractable to medical management. These can be very challenging. There are numerous methods to treat them; below are some examples. The main risks are failure to improve the pain and recurrent neuroma.

Surgical Techniques (Fig. 24.1)

Resection and Grafting (Fig. 24.2)

- For a neuroma in continuity where there is a proximal and distal nerve accessible, assuming that the nerve is not functional.
- The neuroma is resected down to normal-appearing nerve proximally and distally.
- A bridge graft can be used to bridge the gap between the proximal and distal nerve stumps.
- Autografts can be used but have the risk of developing a neuroma on the donor nerve. Allografts are preferred.

Resection and Tube Repair

- As an alternative to grafting, after neuroma resection, a connecting tube can be used to bridge the gap.
- The proximal and distal nerve stumps are sutured to the ends of the tube.

Resection and Diversion (Fig. 24.3)

- When a distal stump is not available for repair.
- The neuroma is resected down to normal-appearing nerve.
- The new stump needs to be diverted away from the skin incision and the neighboring joints to avoid recurrence of a painful neuroma.
- To achieve this goal, the proximal stump can be buried deep in a muscle [1], another nerve (end to side) [2], bone, or a vascularized fascial flap [3].
- To gain length on the stump, a nerve graft can be used, then the distal end of the graft is buried in a muscle. It is the author's preference to suture the nerve to the muscle and use glue to prevent dislodgement.

Resection and Dispersion (Fig. 24.4)

- This is typically indicated after resecting a neuroma from a large mixed nerve, when there is not enough length to bury the proximal stump deep in a muscle. This is usually the case in amputation stumps after multiple surgical resections.
- The neuroma is first resected.
- The proximal stump is divided into multiple fascicles.
- Several grafts are used to suture to different groups of fascicles.
- The distal ends of the grafts are sutured to different locations deep in a neighboring muscle.
- Allografts are preferred.

Electronic supplementary material The online version of this chapter 10.1007/978-3-319-14520-4_24 contains supplementary material, which is available to authorized users.

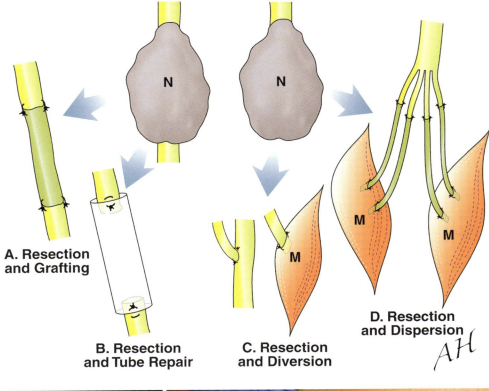

Fig. 24.1 Diagrammatic representation of different techniques to treat a painful neuroma (*N*). If distal normal nerve is available, one may attempt repair with a nerve graft or a tube (options *A*, *B*). If no distal nerve is available, the nerve stump can be buried in muscle (*M*), bone, another nerve (end to side), or a vascularized flap (option *C*). For a large neuroma on a mixed nerve with a short stump (especially amputation stumps), the nerve can be divided into several fascicles, and grafts can be used to gain length and bury the distal end in different muscle (*M*) locations (option *D*)

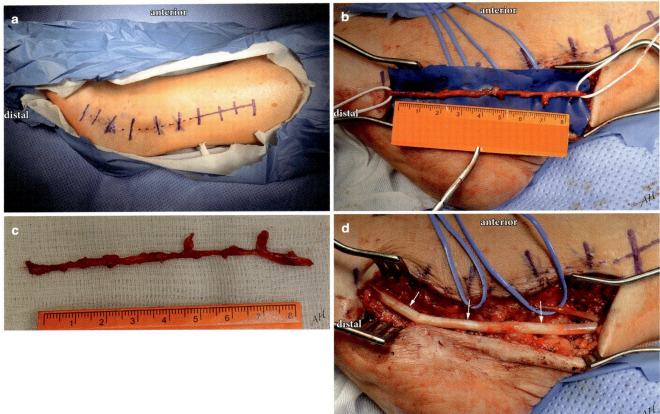

Fig. 24.2 This patient had a painful scarred left sural nerve after ankle surgery, treated by resection and grafting. (**a**) The patient is positioned lateral on a beanbag. (**b**) About 9 cm of sural nerve is exposed, with evidence of scarring and nerve damage. (**c**) The damaged nerve is first resected. (**d**) An allograft (*arrows*) is used to stitch to both proximal and distal nerve stumps. The suture sites are covered with wrap tubes; glue is then applied. Note the short saphenous vein in blue loops

Surgical Techniques

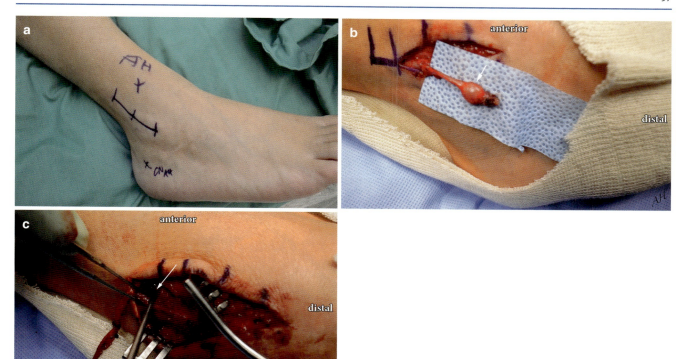

Fig. 24.3 This patient had a painful neuroma along the intermediate dorsal cutaneous nerve. (**a**) She is positioned supine with a hip bump. (**b**) The neuroma is exposed (*arrow*), but no distal stump is found. (**c**) A bed is created in a neighboring muscle (peronei) (*arrow*) to bury the nerve ending after resection of the neuroma

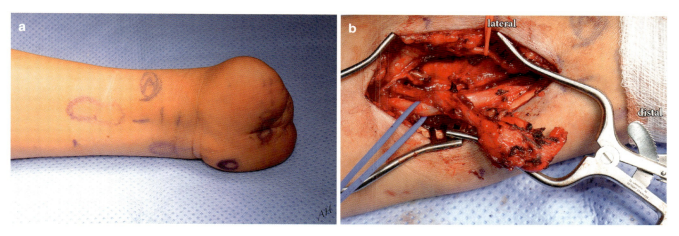

Fig. 24.4 This patient suffered from a severely painful amputation stump. (**a**) Neuroma sites are mapped with ultrasound. (**b**) A neuroma is found at the end of the anterior interosseous nerve (AIN). Note the median nerve in a blue loop and the radial artery in a red loop. (**c**) The neuroma is resected. (**d**) The nerve stump is divided into two, and each part received a piece of allograft (*arrows*). (**e**) The distal ends of the allografts are buried separately in a muscle (*arrows*)

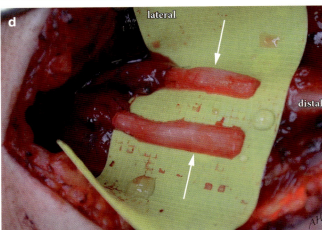
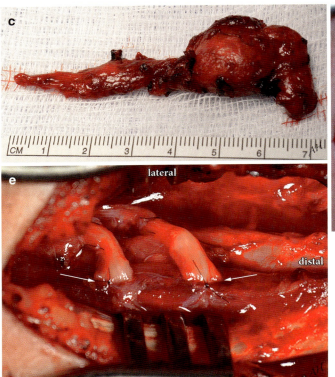

Fig. 24.4 (continued)

References

1. Mackinnon SE, Dellon AL, Hudson AR, Hunter DA (1985) Alteration of neuroma formation by manipulation of its microenvironment. Plast Reconstr Surg 76:345–353
2. Aszmann OC, Moser V, Frey M (2010) Treatment of painful neuromas via end-to-side neurorraphy. Handchir Mikrochir Plast Chir 42(4):225–232
3. Elliot D (2014) Surgical management of painful peripheral nerves. Clin Plast Surg 41:589–613

Nerve Sheath Tumors

Types

Most nerve sheath tumors are benign. They can be schwannomas or neurofibromas. They either occur sporadically or as part of neurofibromatosis (NF) I (neurofibromas) or II (schwannomas), or schwannomatosis. Malignant peripheral nerve sheath tumors (MPNST) are rare but more common in NF I, especially in plexiform neurofibroma. They are suspected based on rapid growth, severe pain, motor deficits, heterogeneous appearance on the MRI, and/or a positive PET (positron emission tomography) scan. Needle biopsy can be performed.

Treatment Strategy

1. Surgery for benign tumors should aim at preserving function. Whenever possible, the parent nerve should be preserved, and only the fascicles harboring the tumor are sacrificed (described below). Small cutaneous nerves can be sacrificed especially with neurofibromas.
2. In neurofibromatosis or schwannomatosis, only symptomatic or fast growing tumors need resection.
3. Dumbbell tumors may require a combined intradural/extradural procedure. The need for spine stabilization depends on the extent of bony resection and the level of the spine.
4. Surgery for MPNST should aim at preserving life. Function can be sacrificed to obtain an oncological resection with negative margins. It requires a team approach with an oncologic surgeon, medical oncologist, radiation oncologist, and a rehabilitation team.

Technique of Removal of a Benign Nerve Sheath Tumor [1]

1. Expose the nerve harboring the tumor based on clinical (Fig. 25.1) and imaging (Fig. 25.2) localization. Intraoperative ultrasound can be used to assist localizing small and deep tumors. Individual nerve exposures are discussed in Parts I and II of this book.

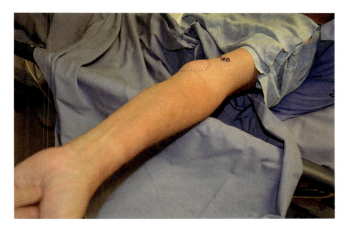

Fig. 25.1 This patient presented with a tender mass (*dotted circle*) in the left arm

---

Electronic supplementary material The online version of this chapter 10.1007/978-3-319-14520-4_25 contains supplementary material, which is available to authorized users.

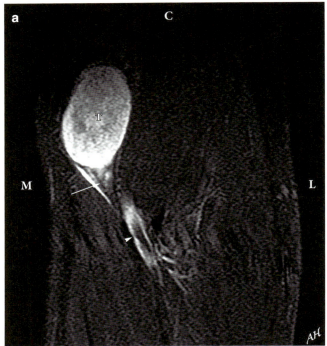

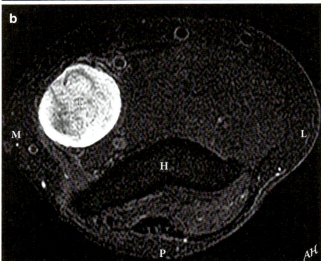

Fig. 25.2 Coronal (**a**), and axial (**b**) T2-weighted STIR MRI of the left arm showing a tumor (*T*) along the median nerve (*arrow*). Note brachial vessels (*arrowhead*). *M* medial, *L* lateral, *C* cranial, *P* posterior, *H* humerus

2. Once the nerve and the tumor are exposed (Fig. 25.3), the tumor surface is mapped with a nerve stimulator to identify a safe entry zone free of fascicles.
3. The nerve sheath is opened using tenotomy or microscissors.
4. Identify the plane between the pseudocapsule and the tumor capsule. This is a safe plane that allows dissection of the tumor with preservation of most bystander fascicles within and outside the pseudocapsule (Fig. 25.4). A spatula or Roton 8 is used for this dissection.

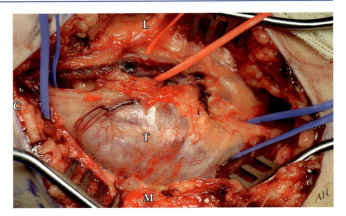

Fig. 25.3 Operative view of the tumor (*T*) within the median nerve (*blue loops*). The brachial vessels are in the red loop. *M* medial, *L* lateral, *C* cranial. Using a nerve stimulator, the tumor is mapped for a safe entry zone

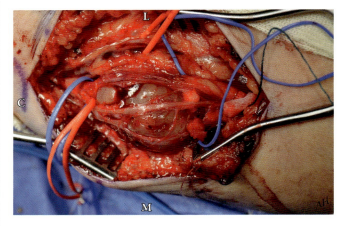

Fig. 25.4 Dissection around the tumor capsule allows preservation of all normal-functioning fascicles. *M* medial, *L* lateral, *C* cranial

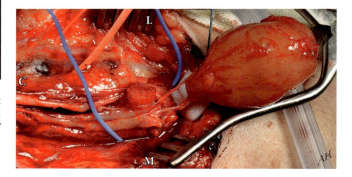

Fig. 25.5 Further dissection reveals the pathological fascicles entering the tumor. These can be coagulated and cut. Stimulation usually reveals nonmotor fascicles. *M* medial, *L* lateral, *C* cranial

5. At the two poles of the tumor, one or more fascicles are seen entering and exiting the tumor (Fig. 25.5). Stimulation is used to confirm these are nonfunctional fascicles. They can then be coagulated with bipolar coagulation and cut, and the tumor can be removed (Figs. 25.6 and 25.7).

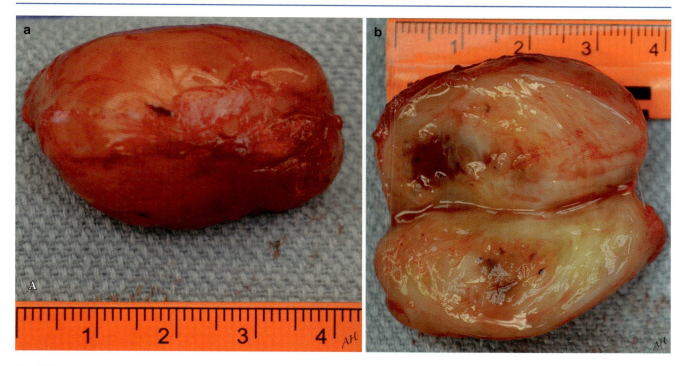

Fig. 25.6 Gross appearance (**a**) and cut surface (**b**) of the tumor

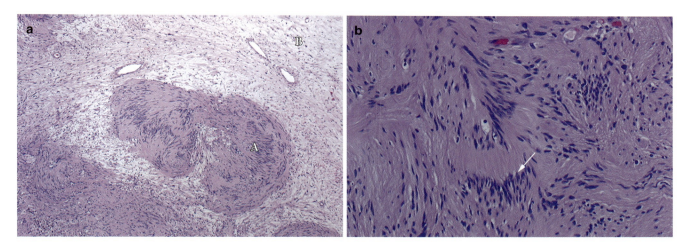

Fig. 25.7 Typical microscopic appearance of a schwannoma. (**a**) 4x magnification, *A* Antoni A, *B* Antoni B. (**b**) 20x magnification of Antoni *A* showing Verocay bodies (*arrow*)

Reference

1. Kim DH, Murovic JA, Tiel RL, Moes G, Kline DG (2005) A series of 397 peripheral nerve sheath tumors: 30-year experience at Louisiana State University Health Sciences Center. J Neurosurg 102:246–255

Intraneural Ganglion Cysts

Mechanism of Formation

Our understanding of intraneural ganglion cysts has considerably improved in the past few years. They usually arise from a degenerated joint. Synovial fluid under pressure exits the joint through an articular branch of a nerve. This can spread retrograde to involve the parent nerve. The classic example is the common peroneal nerve and the superior tibiofibular joint (Fig. 26.1).

Treatment Strategy

To prevent recurrence, all the following should be performed:
1. Decompress the intraneural cyst.
2. Resect or fuse the related joint.
3. Resect the articular branch.

Technique for Peroneal Intraneural Ganglion Cyst [1]

1. The patient is positioned supine with a hip bump to allow access to the anterior and the lateral aspects of the knee joint (Fig. 26.2).
2. The common peroneal (common fibular) nerve and its branches are exposed (Chap. 20) (Fig. 26.3). The incision is extended more anteriorly to access the superior tibiofibular joint (STFJ). Care should be taken when using the scalpel or the Bovie cautery since the nerve can be ballooned and very superficial.

Electronic supplementary material The online version of this chapter 10.1007/978-3-319-14520-4_26 contains supplementary material, which is available to authorized users.

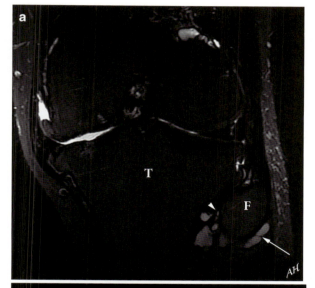

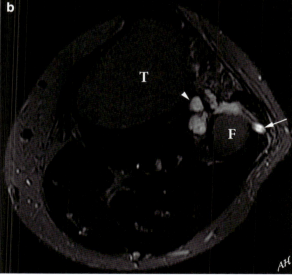

Fig. 26.1 Coronal (**a**) and axial (**b**) T2-weighted fat-saturated MRI of the left knee showing ganglion cyst along the left peroneal nerve (*arrow*) and in the superior tibiofibular joint (STFJ) (*arrowhead*). T tibia, F fibula

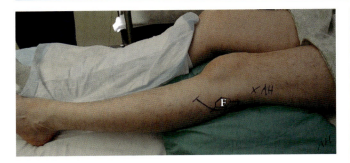

Fig. 26.2 The patient is positioned supine with a bump under the ipsilateral hip. *F* marks the surface anatomy of the fibular head with the insertion of the biceps femoris tendon cranial to it

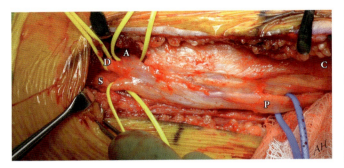

Fig. 26.3 The common peroneal (common fibular) nerve (*P*) is exposed with its terminal branches: superficial (*S*), deep (*D*), and articular (*A*). Note the dilated articular branch and the adjacent peroneal nerve. *C* cranial. The nerve sheath should be opened to drain the cyst

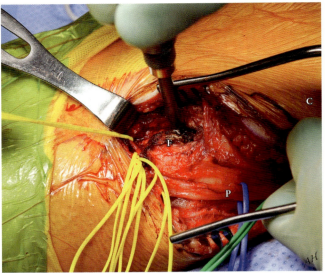

Fig. 26.4 The STFJ is resected using a drill. *F* fibular head, *P* common peroneal nerve, *C* cranial

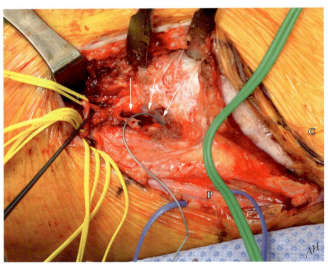

Fig. 26.5 The branch entering the STFJ from the articular branch is identified (nerve hook) and resected. Note that the STFJ has been resected (*arrows*). *P* common peroneal nerve, *C* cranial

3. Once the nerve sheath is opened, pathology can be immediately identified. The nerve is thicker than usual, and the wall can be very thin from cystic dilatation.
4. The intraneural cyst can be decompressed using a no.11 scalpel blade; a slit is made along the nerve axis, and the synovial fluid is expressed out. This allows further dissection without nerve damage.
5. The nerve is followed distally until its trifurcation (superficial peroneal, deep peroneal, and articular branches). The recurrent genicular (articular) branch is then followed until the STFJ. It is important to note that this branch also supplies some anterior compartment muscles and should not be sacrificed in its entirety.
6. An orthopedic surgeon then assists in resection or fusion of the STFJ. Resection is achieved by removal of the articular surface of the fibular head (Fig. 26.4).
7. The branches of the articular branch entering the joint should then be resected (Fig. 26.5).

Reference

1. Spinner RJ, Desy NM, Rock MG, Amrami K (2007) Peroneal intraneural ganglia. Part I. Techniques for successful diagnosis and treatment. Neurosurg Focus 22(6):E16

Peripheral Nerve Stimulation

Indication

Peripheral nerve stimulation is an off-label use of spinal cord stimulators. It is used for intractable pain in a peripheral nerve distribution, refractory to medical management, especially with complex regional pain syndrome (CRPS) (Fig. 27.1) [1, 2].

Surgical Technique

The electrodes can be implanted percutaneously or via an open approach. The open approach is described here.
1. The target nerve is exposed as described in previous chapters (Fig. 27.2).
2. A subcutaneous pocket is created for the implantable pulse generator (IPG) placement either in the infraclavicular area for the upper extremity or in the buttock for the lower extremity.
3. An extension cable is tunneled between the two incisions using a shunt passer (Fig. 27.3).
4. Different electrodes can be used. Our preference is a mesh electrode, which allows suturing to surrounding tissues to prevent migration. The electrode is placed under the nerve and 2-0 Ethibond is usually used to secure it (Fig. 27.4).
5. The electrode is connected to the extension cable and secured using a screwdriver, the sleeve, a suture, and the glue provided.

Electronic supplementary material The online version of this chapter 10.1007/978-3-319-14520-4_27 contains supplementary material, which is available to authorized users.

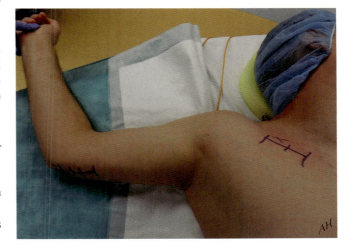

Fig. 27.1 This patient had intractable pain in the ulnar nerve distribution from previous nerve injury. He is positioned supine with the arm abducted for placement of a peripheral nerve stimulator (PNS)

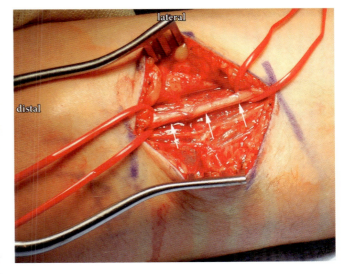

Fig. 27.2 The right ulnar nerve (*arrows*) is exposed in the lower arm

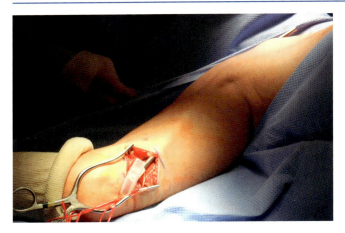

Fig. 27.3 A shunt passer is used to tunnel the extension cable between the two incisions

6. The extension cable is then connected to the IPG and impedance is checked before final placement of the IPG.
7. The IPG is then placed in the subcutaneous pocket created and secured with sutures; we use 2-0 Ethibond (Fig. 27.5).
8. At the proximal and distal sites, extra loops of cable are fashioned to allow for limb mobility without disconnecting the system (Fig. 27.6).

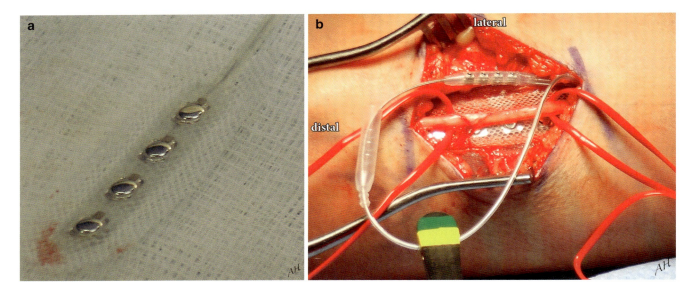

Fig. 27.4 An electrode with a mesh (**a**) is placed under the ulnar nerve and sutured to the underlying tissues (**b**). The electrode is then connected to the extension cable (**b**)

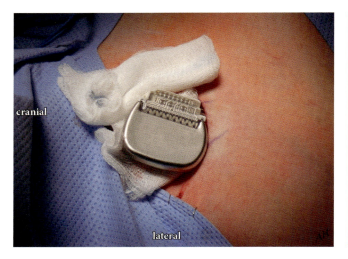

Fig. 27.5 An implantable pulse generator (IPG) is connected to the other end of the connection cable and, once impedance is tested and acceptable, it can be placed in a pocket created in the infraclavicular area

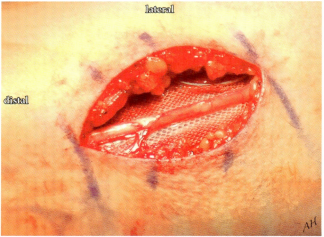

Fig. 27.6 Final position of the electrode under the ulnar nerve, with the mesh sutured to the underlying tissues and a loop of cable secured with sutures. Loops made of extra wire should be fashioned at both incision sites to allow for limb movement without disruption of the system

References

1. Nashold BS Jr, Goldner JL, Mullen JB, Bright DS (1982) Long-term pain control by direct peripheral-nerve stimulation. J Bone Joint Surg 64-A(1):1–10
2. Reverberi C, Dario A, Barolat G, Zuccon G (2014) Using peripheral nerve stimulation (PNS) to treat neuropathic pain: a clinical series. Neuromodulation 17(8):777–783

Dorsal Root Entry Zone (DREZ) Lesion

Indication

This chapter describes dorsal root entry zone (DREZ) lesion in relation to brachial plexus avulsion. Deafferentation pain is difficult to treat and can sometimes be intractable to medical management. DREZ lesion has an 85 % success rate at 2 years in treating neuropathic pain following brachial plexus avulsion (Fig. 28.1) [1].

Technique

1. Position: prone on chest rolls (or Wilson frame), in head pins.
2. Neuromonitoring with somatosensory evoked potentials (SSEPs) and motor evoked potentials (MEPs).
3. The skin is opened at the midline. The paraspinal muscles can be dissected unilaterally for a hemilaminectomy or bilaterally for laminectomy or laminoplasty.
4. We have used the harmonic bone scalpel (Fig. 28.2) to perform bilateral laminectomy, and a plate and screw system for laminoplasty.
5. The dura mater is opened using a no. 15 scalpel blade in the midline or paramedian on the side of the injury. The opening is enlarged using a Woodson tool and a scalpel. The dura is tacked up using 4-0 Nurolon sutures.
6. The arachnoid mater is opened with a nerve hook and microscissors and tacked to the dura using hemoclips.
7. The operative microscope is brought into the field.

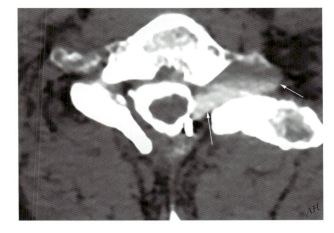

Fig. 28.1 Axial CT myelogram revealing left pseudomeningocele (*arrows*) from brachial plexus avulsion

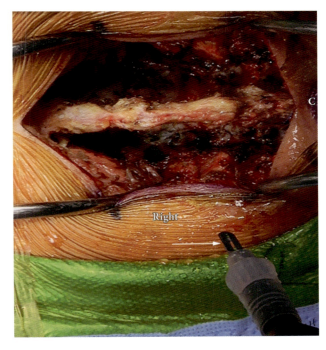

Fig. 28.2 A harmonic bone scalpel (*arrow*) is used to perform a laminectomy at the target levels. *C* cranial

Electronic supplementary material The online version of this chapter 10.1007/978-3-319-14520-4_28 contains supplementary material, which is available to authorized users.

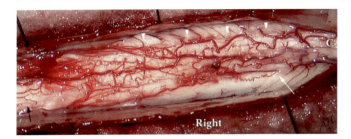

Fig. 28.3 After opening the dura and arachnoid, normal rootlets were observed on the left side (*arrowheads*). The DREZ on the avulsion side is identified by a line joining the last normal rootlet cranial to the injury (*white arrow*) to the first normal rootlet caudal to the injury (*black arrow*). *C* cranial

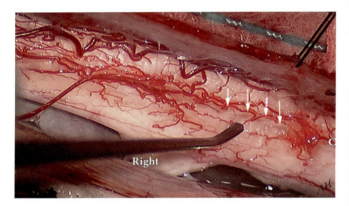

Fig. 28.4 The sites of nerve root avulsion are identified by pits (*arrows*). *C* cranial

8. Identification of the DREZ:
   - The DREZ lines up with an imaginary line joining the last normal rootlet above to the first normal rootlet below the injury (Fig. 28.3).
   - The distance from the DREZ on the avulsed side to the dorsal median sulcus of the cord should be the same as from the latter to the normal dorsal rootlets on the other side. However the cord can be shifted or rotated due to the injury.
   - Pits (the most reliable): these are the sites of avulsion of the dorsal rootlets from the spinal cord (Fig. 28.4). They are defects in the pia mater and are darker in color.
9. The lesion can be done mechanically (Fig. 28.5) [2] or using radiofrequency [3]. Here we describe the mechanical lesion. The lesion should include the avulsed segment with or without the first normal root above and below.
10. Bipolar coagulation with fine tips is used to coagulate the vessels and pia in the DREZ.
11. A myelotomy knife is used to create skip lesions in the DREZ, at about 30–45° angle to the spinal cord surface, to a depth of 2–3 mm. The lesions are 1.5–2 cm in length, leaving 1–2 mm of intact pia in between (Fig. 28.6).

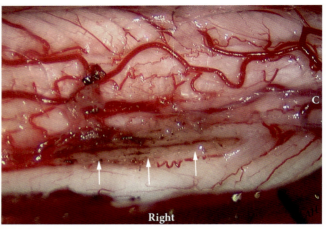

Fig. 28.5 Under the operative microscope, the DREZ lesion is performed (*arrows*) using a myelotomy knife and bipolar coagulation. Angle: 30–45° with the spinal cord medially. Depth: 2–3 mm. The lesion could also be performed using radiofrequency (RF). *C* cranial

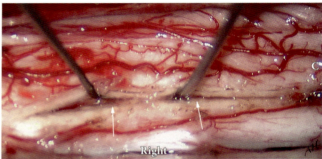

Fig. 28.6 Skip lesions of 1.5–2 cm in length are performed (*arrows*), leaving 1–2 mm of intact pia in between. The latter is then coagulated. *C* cranial

Fig. 28.7 Overview of the end of the procedure showing the entire extent of a right DREZ lesion. *C* cranial

12. Bipolar coagulation is used again to coagulate the walls of the lesion and the skip areas (Fig. 28.7).
13. If normal rootlets above and below are to be included, these are lifted up by an assistant using a nerve hook, and the lesion is created as above (Fig. 28.8).
14. Hemostasis is performed, and then copious irrigation is used to wash out any blood and replenish the cerebrospinal fluid (CSF).

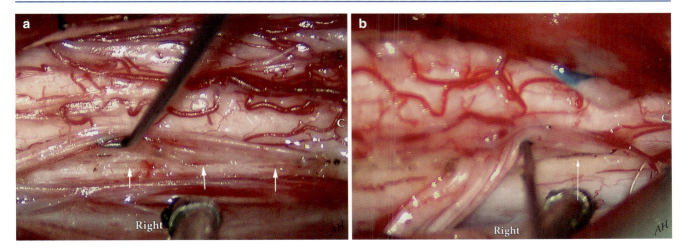

Fig. 28.8 The DREZ (*arrows*) could also include the nerve root caudal (**a**) and cranial (**b**) to the avulsion. *C* cranial

Fig. 28.9 The laminae are replaced with miniplates and screws. *C* cranial

15. The dura is closed using 6-0 Gore-Tex, 5-0 Prolene, or 4-0 Nurolon.
16. This is secured by fibrin or hydrogel glue.
17. The laminae are replaced with miniplates and screws in case of laminoplasty (Fig. 28.9).
18. Careful hemostasis of the muscle. No drain is left. Vancomycin powder can be used in the muscle and subcutaneous layer. Local anesthetic could be injected in the muscle and subcutaneous layers.
19. The fascia is closed tightly with 0 Vicryl sutures.
20. The subcutaneous tissue is approximated with 2-0 Vicryl.
21. The skin is closed with running 3-0 Nylon.

Complications

1. Failure to treat the pain.
2. Recurrent pain.
3. Loss of SSEPs and/or MEPs. Troubleshoot: check the connections, raise the blood pressure, warm irrigation to the spinal cord, and wait a few minutes. If significant changes persist, may perform a wake up test or abort the procedure.
4. Spinal cord injury.
5. CSF leak. This can be treated by elevating the head of the bed, oversewing the incision, or inserting a lumbar drain. If unsuccessful, it will require wound revision.
6. Wound infection.
7. Postoperative hematoma.

References

1. Cetas JS, Saedi T, Burchiel KJ (2008) Destructive procedures for the treatment of nonmalignant pain: a structured literature review. J Neurosurg 109:389–404
2. Sindou M, Mifsud JJ, Boisson D, Goutelle A (1986) Selective posterior rhizotomy in the dorsal root entry zone for treatment of hyperspasticity and pain in the hemiplegic upper limb. Neurosurgery 18(5):587–595
3. Friedman AH, Nashold BS Jr, Bronec PR (1988) Dorsal root entry zone lesions for the treatment of brachial plexus avulsion injuries: a follow-up study. Neurosurgery 22(2):369–373

Part V
Axial Cuts

Upper Extremity Axial Cuts

29

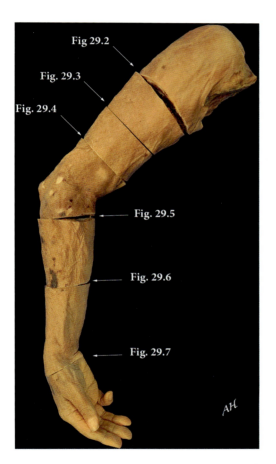

Fig. 29.1 Overview of the right arm showing the location of the axial cuts depicted in the following figures

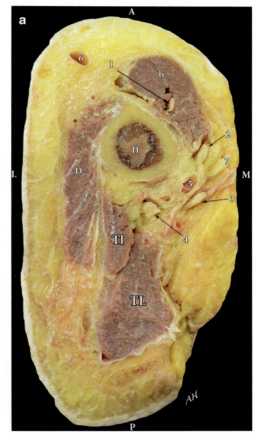

Fig. 29.2 (**a**) Axial cut between the upper and middle thirds of the right arm. *A* anterior, *P* posterior, *M* medial, *L* lateral, *H* humerus, *D* deltoid, *b* biceps brachii, *Tl* triceps, lateral head, *TL* triceps, long head, *1* musculocutaneous nerve, *2* median nerve, *3* ulnar nerve, *4* radial nerve, *5* brachial artery and veins, *6* cephalic vein, *7* medial antebrachial cutaneous nerve. (**b**) Axial T1-weighted MRI at a similar level. *White arrow* radial nerve, *white arrowhead* musculocutaneous nerve, *black arrow* median nerve, *black arrowhead* medial antebrachial cutaneous nerve, *double white arrows* ulnar nerve

A.S. Hanna, *Anatomy and Exposures of Spinal Nerves*,
DOI 10.1007/978-3-319-14520-4_29, © Springer International Publishing Switzerland 2015

Fig. 29.2 (continued)

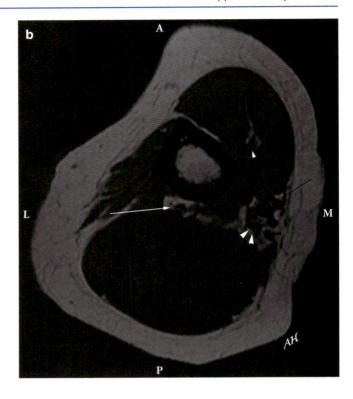

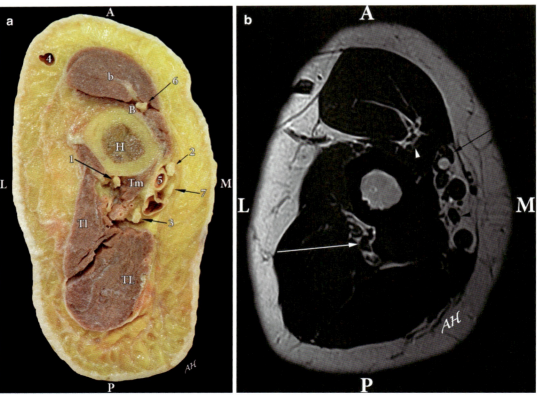

Fig. 29.3 (**a**) Axial cut at the midarm. *A* anterior, *P* posterior, *M* medial, *L* lateral, *H* humerus, *B* brachialis, *b* biceps brachii, *Tm* triceps, medial head, *Tl* triceps, lateral head, *TL* triceps, long head, *1* radial nerve in the spiral groove, *2* median nerve, *3* ulnar nerve, *4* cephalic vein, *5* brachial artery, *6* musculocutaneous nerve, *7* medial antebrachial cutaneous nerve. (**b**) Axial T1-weighted MRI at a similar level. *A* anterior, *P* posterior, *M* medial, *L* lateral, *white arrow* radial nerve in the spiral groove, *white arrowhead* musculocutaneous nerve, *black arrow* median nerve, *black arrowhead* ulnar nerve

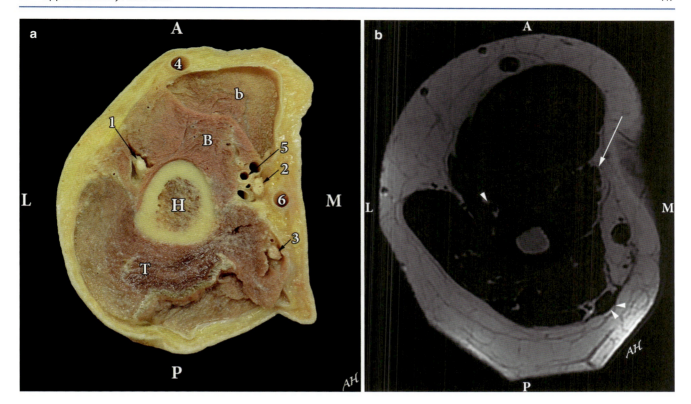

Fig. 29.4 (a) Axial cut between the middle and lower thirds of the right arm. *A* anterior, *P* posterior, *M* medial, *L* lateral, *H* humerus, *B* brachialis, *b* biceps brachii, *T* triceps, *1* radial nerve, *2* median nerve, *3* ulnar nerve, *4* cephalic vein, *5* brachial artery and veins, *6* basilic vein. (b) Axial T1-weighted MRI at a similar level. *Arrow* median nerve, *double arrowheads* ulnar nerve (swollen), *arrowhead* radial nerve

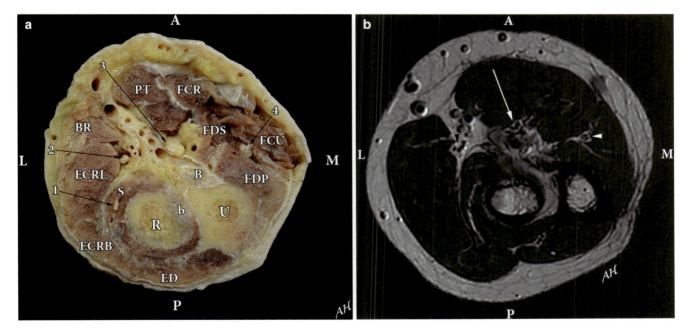

Fig. 29.5 (a) Axial cut just distal to the right elbow. *A* anterior, *P* posterior, *M* medial, *L* lateral, *R* radius, *U* ulna, *BR* brachioradialis, *ECRL* extensor carpi radialis longus, *ECRB* extensor carpi radialis brevis, *ED* extensor digitorum, *S* supinator, *PT* pronator teres, *FCR* flexor carpi radialis, *FDS* flexor digitorum superficialis (sublimis), *FCU* flexor carpi ulnaris, *FDP* flexor digitorum profundus, *B* brachialis, *b* biceps brachii, *1* posterior interosseous nerve (PIN), *2* superficial radial nerve, *3* median nerve, *4* ulnar nerve. (b) Axial T1-weighted MRI at a similar level. *White arrow* median nerve, *white arrowhead* ulnar nerve, *black arrowhead* radial nerve

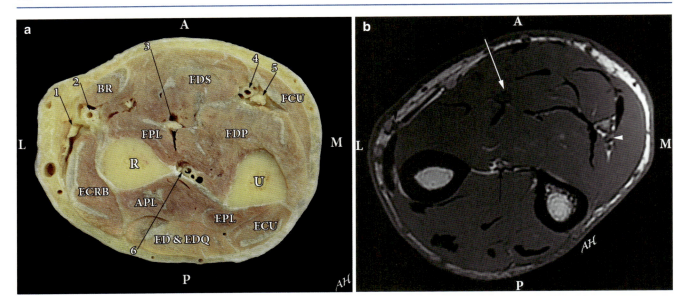

Fig. 29.6 (**a**) Axial cut through the right mid-forearm. *A* anterior, *P* posterior, *M* medial, *L* lateral, *R* radius, *U* ulna, *BR* brachioradialis, *FDS* flexor digitorum superficialis (sublimis), *FCU* flexor carpi ulnaris, *FPL* flexor pollicis longus, *FDP* flexor digitorum profundus, *ECRB* extensor carpi radialis brevis, *APL* abductor pollicis longus, *EPL* extensor pollicis longus, *ECU* extensor carpi ulnaris, *ED & EDQ* extensor digitorum and extensor digiti quinti (minimi), *1* superficial radial nerve, *2* radial artery, *3* median nerve, *4* ulnar artery, *5* ulnar nerve, *6* anterior interosseous nerve (AIN) and vessels. (**b**) Axial T1-weighted MRI at a similar level. *White arrow* median nerve, *white arrowhead* ulnar nerve, *black arrow* anterior interosseous nerve and vessels, *black arrowhead* superficial radial nerve

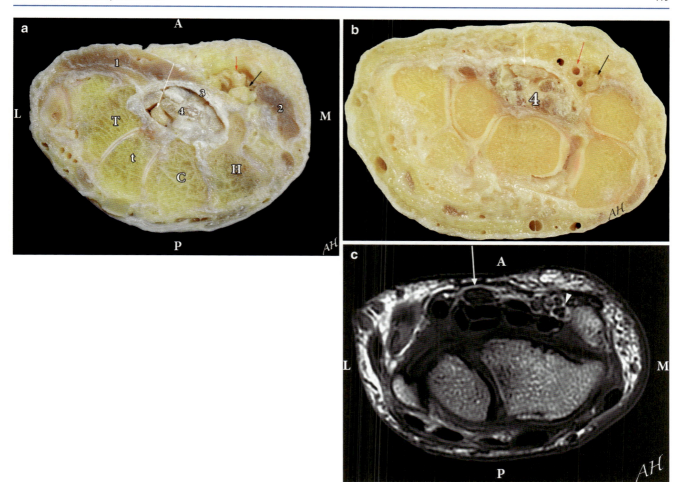

Fig. 29.7 (a) Axial cut through the distal wrist bones of the right side: *A* anterior, *P* posterior, *M* medial, *L* lateral, *T* trapezium, *t* trapezoid, *C* capitate, *H* hamate, *1* thenar muscles, *2* hypothenar muscles, *3* transverse carpal ligament, *4* flexor tendons, *white arrow* median nerve, in this specimen abnormally deep to the flexor tendons, *black arrow* ulnar nerve, *red arrow* ulnar artery. (b) Similar to figure **a**, but the median nerve (*white arrow*) is in its normal anatomical location superficial to the flexor tendons (4). (c) Axial T1-weighted MRI of the right wrist. *Arrow* median nerve in carpal tunnel, *arrowhead* ulnar nerve in Guyon's canal

Lower Extremity Axial Cuts

30

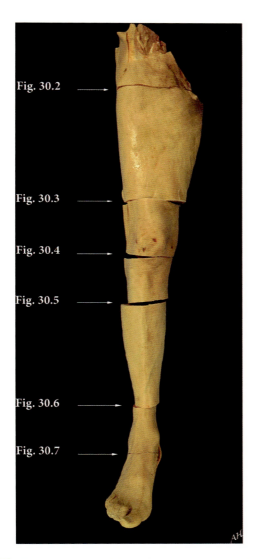

Fig. 30.1 Overview of the right leg showing the location of the axial cuts depicted in the following figures

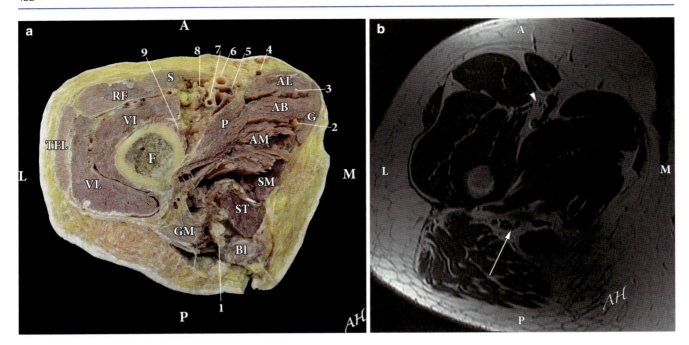

Fig. 30.2 (a) Axial cut through the right femoral triangle. *A* anterior, *P* posterior, *M* medial, *L* lateral, *S* sartorius, *RF* rectus femoris, *VI* vastus intermedius, *VL* vastus lateralis, *TFL* tensor fasciae latae, *G* gracilis, *AL* adductor longus, *AB* adductor brevis, *AM* adductor magnus, *P* pectineus, *SM* semimembranosus, *ST* semitendinosus, *Bl* biceps femoris, long head, *GM* gluteus maximus, *1* sciatic nerve, *2* obturator nerve, posterior division, *3* obturator nerve, anterior division, *4* long saphenous vein, *5* femoral vein, *6* femoral artery, *7* profunda femoris artery, *8* femoral nerve, *9* vastus medialis. (b) Axial T1-weighted MRI at a similar level. *Arrow* sciatic nerve. *Arrowhead* femoral nerve branches

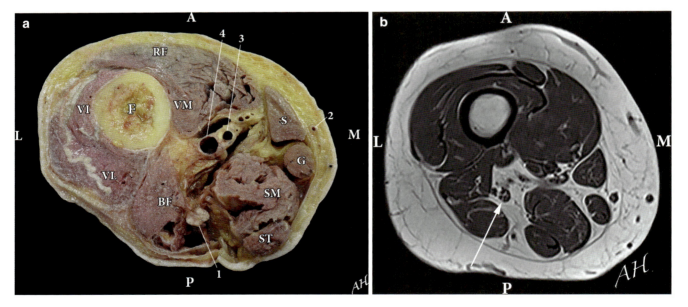

Fig. 30.3 (a) Axial cut through the lower thigh. *A* anterior, *P* posterior, *M* medial, *L* lateral, *S* Sartorius, *G* gracilis, *SM* semimembranosus, *ST* semitendinosus, *BF* biceps femoris, *VL* vastus lateralis, *VI* vastus intermedius, *VM* vastus medialis, *RF* rectus femoris, *1* sciatic nerve, *2* long saphenous vein, *3* popliteal artery, *4* popliteal vein. (b) Axial T1-weighted MRI at a similar level. *Arrow* sciatic nerve

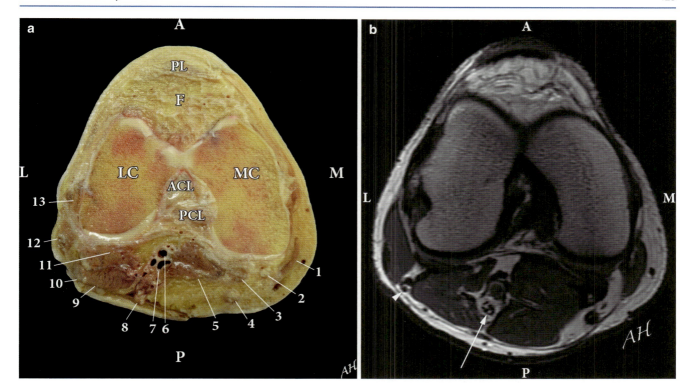

Fig. 30.4 (a) Axial cut through the right knee joint. *A* anterior, *P* posterior, *M* medial, *L* lateral, *MC* medial femoral condyle, *LC* lateral femoral condyle, *ACL* anterior cruciate ligament, *PCL* posterior cruciate ligament, *F* infrapatellar pad of fat, *PL* patellar ligament, *1* sartorius, *2* gracilis, *3* semimembranosus, *4* semitendinosus, *5* gastrocnemius, medial head, *6* popliteal artery, *7* popliteal vein, *8* tibial nerve, *9* gastrocnemius, lateral head, *10* common peroneal (common fibular) nerve, *11* plantaris, *12* biceps femoris, *13* popliteus. (b) Axial T1-weighted MRI at the knee joint. *Arrow*, tibial nerve; *arrowhead*, common peroneal nerve

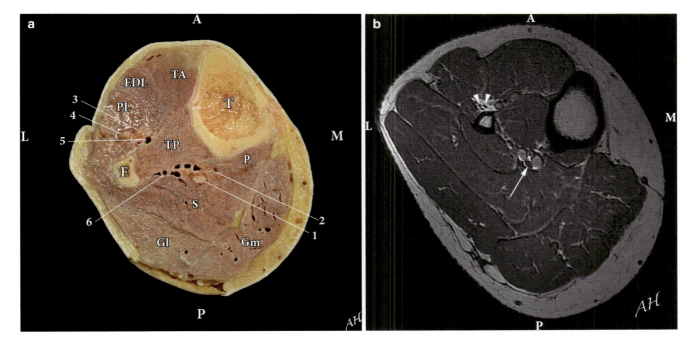

Fig. 30.5 (a) Axial cut distal to the right tibial tuberosity. *A* anterior, *P* posterior, *M* medial, *L* lateral, *T* tibia, *F* fibula, *TA* tibialis anterior, *EDL* extensor digitorum longus, *PL* peroneus longus, *TP* tibialis posterior, *P* popliteus, *S* soleus, *Gl* gastrocnemius, lateral head, *Gm* gastrocnemius, medial head, *1* tibial nerve, *2* posterior tibial vessels, *3* deep peroneal nerve, *4* superficial peroneal nerve, *5* anterior tibial vessels, *6* peroneal vessels. (b) Axial T1-weighted MRI at a similar level. *Arrow*, tibial nerve; *arrowheads*, from lateral to medial, superficial and deep peroneal nerves, respectively

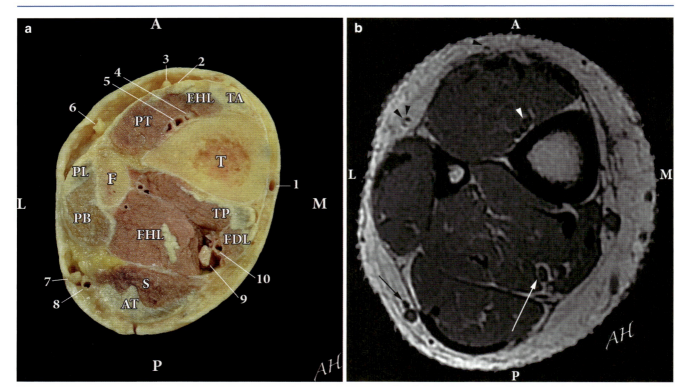

Fig. 30.6 (a) Axial cut proximal to the right malleoli. *A* anterior, *P* posterior, *M* medial, *L* lateral, *T* tibia, *F* fibula, *TA* tibialis anterior, *EHL* extensor hallucis longus, *PT* peroneus tertius, *TP* tibialis posterior, *FDL* flexor digitorum longus, *FHL* flexor hallucis longus, *S* soleus, *AT* Achilles tendon, *PL* peroneus longus, *PB* peroneus brevis, *1* long saphenous vein, *2* extensor digitorum longus, *3* medial dorsal cutaneous nerve, *4* deep peroneal (deep fibular) nerve, *5* anterior tibial vessels, *6* intermediate dorsal cutaneous nerve, *7* sural nerve, *8* short saphenous vein, *9* tibial nerve, *10* posterior tibial artery. *3* and *6* are branches of the superficial peroneal (superficial fibular) nerve. (b) Axial T1-weighted MRI at a similar level. *White arrow*, tibial nerve; *white arrowhead*, deep peroneal nerve; *black arrow*, sural nerve, anterior to the short saphenous vein; single *black arrowhead*, medial dorsal cutaneous nerve; *double black arrowheads*, intermediate dorsal cutaneous nerve

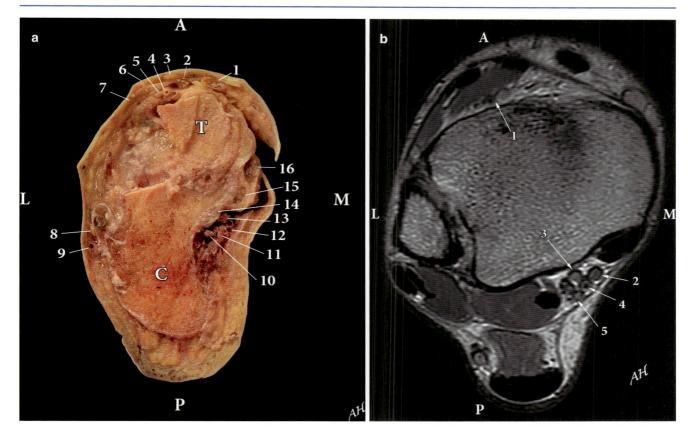

Fig. 30.7 (**a**) Axial oblique cut at the level of the right ankle joint. *A* anterior, *P* posterior, *M* medial, *L* lateral, *C* calcaneus, *T* talus, *1* tibialis anterior, *2* extensor hallucis longus, *3* medial dorsal cutaneous nerve, *4* anterior tibial artery and veins, *5* deep peroneal nerve, *6* intermediate dorsal cutaneous nerve, *7* extensor digitorum longus, *8* sural nerve, *9* short saphenous vein, *10* lateral plantar nerve, *11* calcaneal branch, *12* posterior tibial vessels, *13* medial plantar nerve, *14* flexor hallucis longus, *15* flexor digitorum longus, *16* tibialis posterior. (**b**) Axial cut just proximal to the right ankle joint, in a proton density MRI. *1* deep peroneal nerve and anterior tibial vessels, *2* medial plantar nerve, *3* lateral plantar nerve, *4* posterior tibial vessels, *5* calcaneal branch

Index

A
Abductor digiti minimi, 35
Abductor hallucis, 71
Abductor pollicis brevis, 31
Accessory obturator nerve, 55
Achilles tendon, 77
Adductor brevis, 81
Adductor canal, 59, 60
Adductor longus, 81
Adductor magnus, 67, 68, 81
Adductor pollicis, 35, 36
AIN. See Anterior interosseous nerve (AIN)
Allografts, 89, 95
Anatomical snuffbox, 25
Anconeus epitrochlearis, 36
Anterior division, 15, 19, 23, 55, 67, 81, 90
Anterior interosseous nerve (AIN), 31, 32
Anterior superior iliac spine (ASIS), 59, 63
Arcade of Fröhse, 25
Arcade of Struthers, 36
Articular, 76, 77, 81, 103, 104
ASIS. See Anterior superior iliac spine (ASIS)
Avulsion, 109, 110
Avulsion, 29
Axillary artery, 15, 23, 25, 29, 35
Axillary nerve, 23–24, 51
Axonotmesis, 89

B
Berretini, 31
Biceps brachii, 15, 25, 26, 29, 30, 32, 36, 41, 90
Biceps femoris, 67, 68, 76
Bicipital aponeurosis, 31, 32
Brachial artery, 29, 32, 35, 36
Brachialis, 26, 29–32, 41, 90
Brachial plexus, 11–12, 15–22
Brachioradialis, 25, 26

C
Calcaneal branch, 71, 72
Camper's, 85
Carpal tunnel, 31, 33, 34
Cephalic vein, 15, 26
Clavicle, 11, 19, 45
Common peroneal nerve, 55, 67, 76
Complex regional pain syndrome (CRPS), 105
Coracobrachialis, 15, 29, 30
Coracoid process, 15, 23, 25, 29
CRPS. See Complex regional pain syndrome (CRPS)
Cubital fossa, 31, 32

D
Deep circumflex iliac vessels, 85
Deep peroneal, 76, 77
Deltoid, 15, 23, 24, 26, 41
Deltopectoral groove, 15
Dentate ligament, 45
Dorsal cutaneous branch, 35
Dorsal median sulcus, 110
Dorsal root entry zone (DREZ), 109–111
Dorsal scapular nerve, 11

E
End to end, 91
End to side, 91, 95
Extensor carpi radialis, 25
Extensor digitorum brevis, 76
Extensor digitorum longus, 76, 77
Extensor hallucis longus, 76, 77
Extensor retinaculum, 76, 77
External oblique, 56, 59, 85

F
Fascicle, 67, 90, 91, 99
Fat pad, 11, 12, 45, 67, 68
Femoral artery, 59, 60
Femoral nerve, 55, 56, 59, 63, 81
Femoral sheath, 59
Femoral triangle, 59
Fibular head, 76, 104
Fibular neck, 76
Finger drop, 25
Flexor carpi radialis, 31
Flexor carpi ulnaris, 32, 35, 37
Flexor digiti minimi, 35
Flexor digitorum profundus, 31, 35, 36
Flexor digitorum superficialis, 31, 32
Flexor pollicis brevis, 31, 35
Flexor pollicis longus, 31
Flexor retinaculum, 71
Foot drop, 75
Fröhse arcade, 25, 26
Froment's sign, 36

G
Gastrocnemius, 71
Gemellus inferior, 67, 68
Gemellus superior, 67, 68
Genitofemoral nerve, 55
Gluteus maximus, 55, 67, 68

Gracilis, 51, 81
Grafting, 89–90, 95
Great auricular nerve, 11, 45
Guyon's canal, 35, 37

H
Harmonic bone scalpel, 109
Humerus, 23, 32
Hypothenar, 35–37

I
Iliohypogastric, 55, 56, 85
Ilioinguinal, 55, 56
Ilioinguinal nerve, 85–86
Implantable pulse generator (IPG), 105, 106
Inferior gluteal nerve, 55
Infraclavicular exposure, 15–19, 23, 25, 29
Infraspinatus, 41
Inguinal ligament, 59, 63, 85
Intercostal muscles, 51
Intercostal nerves, 51
Intercostobrachial nerve, 51
Intermediate dorsal cutaneous nerve, 76
Intermetatarsal ligament, 72
Internal oblique, 56, 59, 85
Interossei, 36
Intraneural ganglion cysts, 76, 103
IPG. *See* Implantable pulse generator (IPG)

L
Lacertus fibrosus, 31
Laminectomy, 109
Laminoplasty, 109, 111
Lateral antebrachial cutaneous nerve, 29, 30
Lateral cord, 15, 19, 29
Lateral femoral cutaneous nerve (LFCN), 55, 59, 63, 85
Lateral pectoral nerves, 15
Lateral sural cutaneous nerve, 76
Latissimus dorsi, 15, 49
Leash of Henry, 25, 26
Levator scapulae, 11, 21
LFCN. *See* Lateral femoral cutaneous nerve (LFCN)
Long saphenous vein, 59, 60
Long thoracic nerve, 11, 21, 49
Lower extremity axial cuts, 121–125
Lower triangular space, 25
Lower trunk, 11, 15, 19, 21, 51
Lumbar plexus, 55
Lumbosacral trunk, 55, 56
Lumbrical, 31

M
Malignant peripheral nerve sheath tumors (MPNST), 99
Marinacci, 31
Martin-Gruber, 31
Medial antebrachial cutaneous nerve, 15
Medial brachial cutaneous nerve, 15
Medial cord, 15, 19, 35, 36
Medial dorsal cutaneous, 76
Medial epicondyle, 32, 36
Medial intermuscular septum, 35–37
Medial pectoral nerve, 15

Medial plantar nerve, 71
Medial sural cutaneous, 71
Median nerve, 15, 31–34, 36, 51
Meralgia paresthetica, 63
Mesh electrode, 105
Middle trunk, 11
Morton's neuroma, 71, 72
MPNST. *See* Malignant peripheral nerve sheath tumors (MPNST)
Musculocutaneous nerve, 15, 23, 29, 30, 51
Myelotomy, 110

N
Nerve sheath tumors, 99
Nerve transfers, 90–91
Nervi erigentes, 55
Neurapraxia, 89
Neurofibromatosis (NF), 99
Neuroma, 59, 72, 77, 90, 95
Neuromonitoring, 109
Neurotmesis, 89
NF. *See* Neurofibromatosis (NF)
NF I, 99
NF II, 99

O
Oberlin's, 90
Obturator canal, 81
Obturator externus, 81
Obturator foramen, 81
Obturator internus, 67, 68
Obturator nerve, 55, 56, 81
Olecranon, 36
Omohyoid, 11
Opponens digiti minimi, 35
Opponens pollicis, 31
Osborne's band, 36

P
Palmar aponeurosis, 33
Palmar cutaneous branch, 31, 33–35
Palmaris longus, 33
Palmar recurrent, 31, 33, 34
Parsonage-Turner syndrome, 49
Pectineus, 81
Pectoralis major, 15, 29, 32, 36, 51
Pectoralis minor, 15
Pelvic splanchnic nerve, 55
Peripheral nerve stimulation, 105–106
Peroneus brevis, 76
Peroneus longus, 76
Peroneus tertius, 76
Phrenic nerve, 11, 12
PIN. *See* Posterior interosseous nerve (PIN)
Piriformis, 55, 56, 67, 68
Pits, 110
Platysma, 11
Pleura, 12, 21, 51, 52
Plexiform neurofibrom, 99
Popliteal fossa, 71, 76, 77
Posterior circumflex humeral vessels, 23
Posterior cord, 15, 19, 23, 25, 49
Posterior cutaneous nerve of the thigh, 55, 67
Posterior division, 23, 24, 41, 49, 55, 59, 67, 81

Posterior interosseous nerve (PIN), 25, 26
Posterior triangle of the neck, 11, 45
Profunda brachii, 25
Pronator quadratus, 31
Pronator teres, 31, 32
Psoas, 55, 56, 59, 63, 81, 85
Pubic tubercle, 81
Pudendal nerve, 55, 67, 68

Q
Quadrangular space, 23, 24
Quadratus femoris, 67, 68
Quadriceps, 59

R
Radial nerve, 23, 25–27, 36, 89
Radiofrequency, 110
Recurrent genicular, 76
Regeneration, 91
Repair, 89, 91, 95
Rhomboids, 11, 21
Rib, 21, 22, 51
Riche-Cannieu, 31

S
Sacral plexus, 55, 56
Saphenous nerve, 59, 60, 72, 89
Sartorius, 59, 60, 63
Scalenus anterior, 11
Scalenus medius, 11, 21, 49
Scalenus posterior, 21
Scapula, 21, 41
Scarpa's, 85
Schwannomatosis, 99
Sciatic nerve, 55, 56, 67, 68, 71, 76
Seddon, 89
Semimembranosus, 67, 68
Semitendinosus, 67, 68
Serratus anterior, 11, 21, 49
Serratus posterior superior, 21
Short saphenous vein, 77
Soleal arch, 71
Spinal accessory nerve, 11, 22, 45, 90
Spinal nerves, 11
Spinoglenoid notch, 41
Spiral groove, 25
Sternocleidomastoid (SCM), 11, 45
STFJ. See Superior tibiofibular joint (STFJ)
Stimulation, 11, 45, 52, 57, 67, 81, 91
Struthers' ligament, 32
Subclavian artery, 11, 12
Subclavius, nerve to, 11
Subscapularis, 15, 23
Subscapular nerves, 15
Sunderland, 89
Superficial palmar arch, 33, 34
Superficial peroneal, 76, 77
Superficial radial nerve, 25
Superior gluteal nerve, 55
Superior tibiofibular joint (STFJ), 76, 103
Supinator, 25, 26
Supraclavicular exposure, 11–12
Suprascapular artery, 41

Suprascapular ligament, 41
Suprascapular nerve, 11, 12, 21, 41–44, 90
Suprascapular notch, 21, 41
Suprascapular vessels, 11, 12
Supraspinatus, 41
Sural nerve, 71, 77, 89

T
Tarsal tunnel, 71, 72, 76, 77
Teres minor, 23
Thoracic duct, 12
Thoracodorsal nerve, 15, 49
Tibialis anterior, 76, 77
Tibial nerve, 55, 67, 71
Transposition, 36, 37, 63
Transverse carpal ligament, 31, 33
Transverse cervical vessels, 11, 12
Transversus abdominis, 56, 59, 85
Trapezius, 11, 12, 21, 22, 41, 45
Trauma, 89–94
Triangular interval, 25
Triceps, 23, 25, 26, 36
Tube repair, 91, 95

U
Ulnar artery, 35
Ulnar nerve, 15, 32, 35–40
Ultrasound, 41, 63, 77, 85, 99
Upper lateral cutaneous nerve of the arm, 23
Upper trunk, 11, 21, 23, 29, 41

W
Wartenberg's sign, 36
Wiltse approach, 57
Winging, 45, 49
Wrist drop, 25

Printing: Ten Brink, Meppel, The Netherlands
Binding: PHOENIX PRINT, Würzburg, Germany